Electrospinning of Nanofibers for Battery Applications

Shengjie Peng · P. Robert Ilango

Electrospinning of Nanofibers for Battery Applications

Springer

Shengjie Peng
Department of Applied Chemistry
Nanjing University of Aeronautics
and Astronautics
Nanjing, China

P. Robert Ilango
Department of Applied Chemistry
Nanjing University of Aeronautics
and Astronautics
Nanjing, China

ISBN 978-981-15-1430-2   ISBN 978-981-15-1428-9   (eBook)
https://doi.org/10.1007/978-981-15-1428-9

This Springer imprint is published by the registered company Springer Nature Singapore Pte Ltd.
The registered company address is: 152 Beach Road, #21-01/04 Gateway East, Singapore 189721, Singapore

Preface

At present, nanotechnology is offering new results and prospects to guarantee viable energy for the future on account of their unusual shapes, compositions, and physical/chemical properties. Remarkably, one-dimensional (1D) nanostructures, such as nanofibers, nanorods, and nanotubes, have drawn significant interest because of confinement effects and the structure-correlated properties. One-dimensional nanostructured material can be synthesized by using various techniques, including top-down synthesis and bottom-up method. Among them, electrospinning is an uncomplicated and versatile system which is used in generating 1D nanostructures. The prepared nanofibers can be scaled up for assembly on an industrial scale with tunable resulting characteristics like morphology and diameter.

Interestingly, electrospun-based nanofibers deliver higher electrochemical performance for batteries, fuel cells, and supercapacitors. Though many reports have been published over the years on the electrospun nanofiber-based materials synthesis and their potential application, still the recent development of different nanofiber-based materials for battery applications is not overviewed. Therefore, in this book, we discussed hot and new trending growth of various materials such as metal oxides, metals, and carbon-based composite nanofibers using electrospinning method. Undoubtedly, this book provides more achievement on materials perspectives for their future research and developments in the energy storage systems.

Nanjing, China
Prof. Shengjie Peng
Dr. P. Robert Ilango

Contents

About the Authors

Prof. Shengjie Peng received his Ph.D. degree at Nankai University (P. R. China) in 2010. Following a postdoctoral fellowship with Prof. Alex Yan and Prof. Seeram Ramakrishna at Nanyang Technological University and National University of Singapore, he is now working as a professor at Nanjing University of Aeronautics and Astronautics. He has been working on functional nanomaterials in energy and environment for more than ten years. His current research interests include the design and development of nanomaterials and their applications in energy.

Dr. P. Robert Ilango obtained his Ph.D. degree from Kyung Hee University (South Korea) in 2016. He worked as a postdoc fellow at National Tsing-Hua University (Taiwan, 2016–2018). He also worked as a postdoctoral research fellow at Nanjing University of Aeronautics and Astronautics (China, October 2018–January 2020) under the supervision Prof. Shengjie Peng. Currently, he is working as a postdoctoral research fellow at the University of North Dakota, USA. His research interests focus on the preparation of various nanostructured materials for energy storage applications.

Chapter 1
Electrospinning Technology

Abstract In this chapter, we discuss the instrumentation and underlying principle of electrospinning techniques. Even though there are many methods of fabricating nanofibers, electrospinning is conceivably the most versatile process. Materials such as polymer, composites, ceramic and metal nanofibers have been built using electrospinning directly or through post-spinning processes. We arranged the different types of electrospun nanofibers and their recent development including aligned, random, and core–shell nanofibers. Finally, we reviewed the possible potential applications such as tissue engineering, water treatment and energy storage applications. This analysis may bring overall ideas related to electrospun preparation and their significant applications with the actual mechanism.

Keywords Electrospinning · Types of fiber · Potential applications

1.1 Introduction of Materials

It is well known that the fast population growth and universal energy demand and changes in climate concerns impose a strategy revolution to renewable energy sources. The growth of high-performance energy storage devices has been a gradually high claim to enable resourceful [1, 2]. The extensive application of electric vehicles requires finding substitute power sources with high power/energy densities, long cyclic life and zero-emission. Batteries are the dominant power source to the many portable devices. Carbon materials include carbon nanotube, graphene, and carbon nanofibers have been broadly studied due to their admirable multifunctional activities. Interestingly, they shows good chemical stability, excellent electrical conductivities and large surface areas. Table 1.1 describes various material and their applications [3–5]. Carbon nanofibers have different arrangements subject on the assembling style of graphene layers. The graphene layers can be perpendicular, inclined or even coiled along the fiber axis [6] Amorphous carbon nanofibers could be produced by the spinning of a polymer precursor, accompanied by carbonization at a low-temperature treatment respectively [7]. Because of exceptional properties, the carbon nanofibers have presented great prospective for large-scale applications in many areas, correctly

S. Peng and P. R. Ilango, *Electrospinning of Nanofibers for Battery Applications*, https://doi.org/10.1007/978-981-15-1428-9_1

Table 1.1 Various carbon material and their applications

Carbon derivatives	Properties				Applications	References
	Specific gravity (g c^{-3})	Electrical conductivity (S/cm)	Thermal conductivity (W/(m K))	Surface area (m^2/g)		
Carbon black	0.13–2	5–30	~0.4	10–1443	Conductive additives; electrodes in LABs	[15]
Carbon nanotubes	0.8–1.8	10^2–10^6	2000–6000	50–1315	Conductive additive; electrode materials; substrates for supporting active metal (oxide)	[16]
Carbon nanofibers	1.5–2.0	10^{-7} to 10^3	5–1600 (graphitic CNFs)	20–2500	Conductive additives; electrode materials; substrates for supporting active metal (oxide)	[17]
Graphite	1.9–2.3	4000 (in plane), 3.3 (c-axis)	298^p, 2.2^c	1–20	Anodes in LIB	[16]
Graphene	2.3	106	600–5000	500–2630	Conductive additives; electrode materials; substrates for supporting active metal (oxide)	[18]

energy storage fields. For example, nanofibers with one-dimensional hold a less clumped pattern that has been confirmed to produce enhanced battery characteristics [8, 9]. Chemical vapour deposition (CVD) and spinning methods are used to prepare carbon nanofibers in which CVD method was conducted in early 1980 using hydrocarbons as precursors. Still, this method does not meet cost-effective that takes limited potential applications. In contrast, the spinning method will be conducted by using polymer precursors. There are many spinning techniques such as wet spinning [10], gel spinning [11], dry spinning [12] and melt spinning [13] have been discovered to prepared carbon nanofibers. However, the fibers which prepared from techniques mentioned above had more than 5 μm in diameter that hampered the ions transport to the core level. Later, electrospinning method was demonstrated to prepare carbon nanofibers with low cost in that high voltage is employed to the polymer solution which is expelled in a fiber arrangement creating use of the interface between the charged polymer and external electrical field. It could be possible to prepare <1 μm in diameter with this technology. Hence, the electrons and ions could diffuse into core level of carbon fibers. Furthermore, the electrospinning method can be effortlessly scaled up, hence displaying encouraging imminent for trade manufacture [14].

Herein we have demonstrated the carbon nanofibers and their derivatives towards the battery applications. Table 1.1 illustrates a detailed analysis of open publications related to electrospun carbon nanofiber for battery applications over the past ten years. The massive growth in the number of papers published on various batteries indicates that the materials prepared by the electrospinning process represents fast evolving in this field (Fig. 1.1).

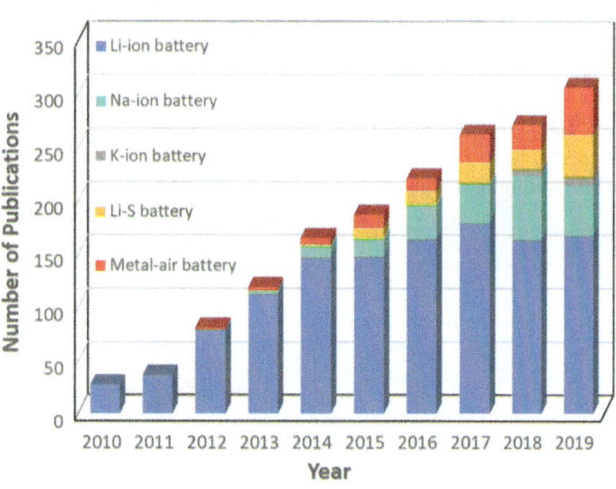

Fig. 1.1 Research articles published on electrospun nanofibers for battery applications from the Web of Science database (**2010–2019**)

1.2 Electrospinning Process

1.2.1 Instrument and Working Principle

A distinctive electrospinning device comprises of the spinneret, high voltage power source and a metal collector [19]. The syringe used to control the flow of polymer solution, spinneret used for ejecting fiber and collected used for collecting fibers [20]. The nozzle diameter approximately 100 μm–1 mm and the air-gap distance between collector. The spinneret is roughly between 5 and 25 cm and high voltage about 100–3000 kV m^{-1} would be given between electrode and collector. A schematic of an electrospinning instrument is displayed in Fig. 1.2.

In a conventional electrospinning experiment, a polymer solution is first pumped over a spinneret to shape a pendant droplet, and a collector is placed across from the spinneret at an adjusted distance. While electrospinning, the precursor solution is ejected from the spinneret, under an electric field, making a small droplet. With an adequately high electric field, the electrostatic forces are significant enough to resolve the polymer solution's surface, forming a liquid jet that travels in the track of the counter electrode. Solvent evaporation of the melt happens through the drive of the liquid jet, depositing, on the collector, the continuous spun fibers commonly as a two-dimensional nonwoven web. Also, the viscosity, solution conductivity, and humidity of the atmosphere impact the subsequent development of the electrospun nanofibers [21]. Besides, different assembly of fibers (random and aligned) can also be fabricated by alteration on the voltage or replace the collecting device. For instance, core–shell and the porous structure can be found using the co-axial nozzle [22], and aligned structures can be attained by rotating collector [23], respectively.

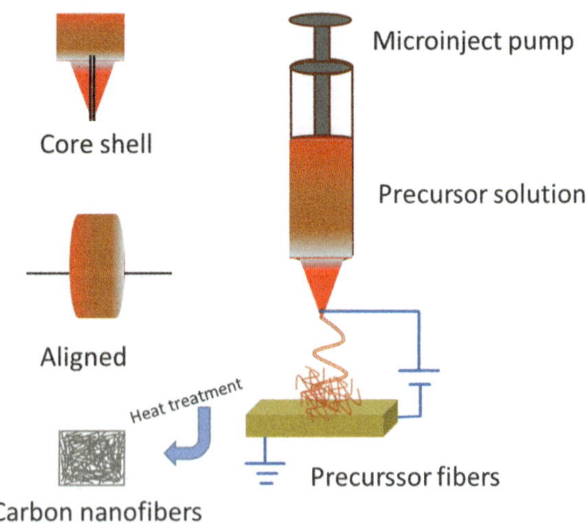

Fig. 1.2 Typical schematic diagram of the electrospinning device

1.2.2 Effect of Precursor

The ideal dealing settings that can effectively yield fibers in electrospinning is power-fully reliant on precursors which further disturb the ultimate fiber structure, such as diameter and length. The main precursors are polymers that are liquefied in a solvent. To understand the extensive usage of different polymer is summarized in Table 1.2. Accordingly, the precursors can be classified into two groups, explicitly host precursor and the guest precursor [20]. The host precursors act as a scaffold for fiber structure, and guest produces void space. Polyacrylonitrile simply PAN is the most common polymer precursor for constructing electrospun CNFs which has a high carbon yield of 50%. Moreover, a two-step thermal treatment is required to convert the spun polymer fibers into CNFs. Also, PAN fibers are thermally stable due to nitrile groups react with adjacent groups to form a ladder structure [24]. There are many interesting polymer precursors have been reported like PVDF [25], PVA [26], PVP [27], PI [28], PS [29], PMMA [30], pitch [31], and lignin [32]. Among them, PVA famous precursor for creating free-standing CNF film. Pitch is another important precursor for the production of CNFs [33]. Impressively, Lignin is a second plentiful polymer in nature which has several aromatic components. It is classified into Al cell and alkali lignin types. Importantly, lignin-based CNFs is has great features of the large specific surface area, ascribed to the activation by high content oxygen that is produced while the stabilization process [34]. In overview of precursors merits and demerits we should consider the crucial facts like yield of carbon and compatibility and instinct properties of precursors which has major effects.

1.2.3 Effect of Parameters

The creation of fibers by the electrospinning method is constructed on the specific progressions and parameters embraced. Most important parameters that affect the structure of electrospun fibers include viscosity, applied electric field and flow rate, spinneret collector, and the type of solvent removal process. The values and signifi-cance of these factors such as viscosity, surface tension, applied electric field, flow rate etc. are extra discussed as follows.

(i) **Viscosity**
Viscosity is the crucial factor in the bead formation and diameter of the fibers which is decided by concentration and molar mass. The Mark-Houwink-Sakurada equation for the linear polymer can be written as

$$\eta = k_1 M^a \tag{1.1}$$

where the constants "k" and "a" relay on the polymer, solvent and temperature [35]. Also, viscosity had a solid influence on fiber diameter. Example, the fiber diameter, d_f, increases with an increase in viscosity that can be written as

Table 1.2 Various polymers and their applications

Precursor	Settings	Functions	Merits	Demerits
PAN	4–10 wt% in DMF 10–30 kV	Carbon source for CNFs and their hybrids	High carbon yield (>50%) and spinnability	Typically only DMF is used as solvent
PVDF	7–25 wt% in DMAc/DMF/acetone, 15–20 kV	Separator and gel electrolyte for LIB	High mechanical properties	Dehydrofluorination for carbonization
PVA	10 wt% in water 8–35 kV	Carbon source for metal oxide/CNF composites	Water soluble	Low carbon yield of 3–10%
PVP	4–10 wt% in ethanol/DMF, 10–25 kV	Carbon source for metal oxide/CNF composites and template for neat metal oxide fibers	Soluble in various solvent	Low carbon yield of ~15%
PI	10–20 wt% in N-methyl, pyrrolidone/THF/methanol/DMAc 8–25 kV	Carbon source for CNFs, separator for LIB	High conductivity of as prepared CNFs	Complex fabrication process
Pitch	30–40 wt%, THF/DMF 18–25 kV	Carbon source for CNFs	High carbon yield (60% at 1000 °C)	Low spinnability
PMMA	8–10 wt% in DMF, 10–20 kV	Scaffold for neat metal oxide fibers and sacrificial phase for voids	Compatible with other host polymer, such as PAN	Difficult to be fully removed, requiring 1000 °C

(continued)

Table 1.2 (continued)

Precursor	Settings	Functions	Merits	Demerits
PS	5–35 wt% in DMF/THF/chloroform, 15–25 kV	Scaffold for neat metal oxide fibers and sacrificial phase for voids	Easy to be decomposed at 450 °C	Poor compatibility with host polymer, beads are easy to form
Lignin	20–35 wt% in DMF/water, 6–26 kV	Carbon source for CNFs	Large surface area	Large diameter; low carbon yield of 20–40%

$$d_f = k_2 \eta^n \tag{1.2}$$

where k_2 is a constant and n depends on the kind of polymer precursor. From the literature, n was originating to be 0.41 and 0.72 for PS and PMMA, respectively [20, 35, 36].

(ii) **Surface tension**

The surface tension of the precursor solution has to be reserved low to diminish the essential applied voltage. The soft and smooth fibers will be formed at low surface tension. Also, the surface tension is fixed by the category of solvent cast-off [37].

(iii) **Applied electric field**

An applied electric field is a vital factor for the electrospinning process. The fiber diameter will decrease concerning electric potential increment.

(iv) **Flow rate**

The feed rate of the precursor solution is another significant aspect that takes on a substantial duty in the jet velocity and the rate of polymer solution deliver [8]. Moreover, it should be noted that the reduction in flowing rate roots to the organization of smaller-diameter fiber, while at high flowing rate fibers produce spindle-like structure [38].

(v) **Others**

Other factors like solvent removal procedure, the geometry of collector, injection setting, electrical conductivity solvent volatility and solvent humidity causes the formation of fiber structure.

1.3   Types of Fibers

To prepare different fiber it requires many alterations in the underlying protocol that are related to engineering progress of spinnerets and collectors for formulating random, aligned, twisted and core/shell nanofibers.

1.3.1   Random Fibers

The trick for fabricating randomly destined fibers is uncomplicated. Hence, a single nozzle with a metal plate or graphite paper collector is used during the preparation process [33, 39]. Also, the collector can be positioned either vertically or horizontally, and the positive and negative electrodes are linked to the spinneret and collector, respectively. Several nozzles are occasionally applied to boost the spinning procedure. According to previous literature, it is reported that low rotation speed <500 used to prepare fiber film [7, 40]. Zhang et al. produced freestanding CNF films for LIBs and the fibers demonstrated in Fig. 1.3.

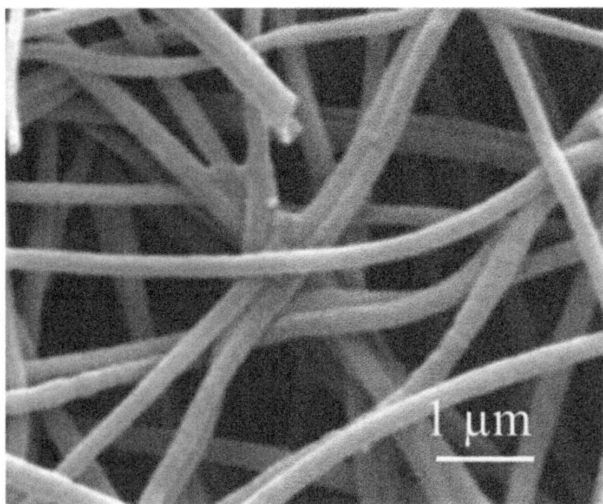

Fig. 1.3 Scanning electron microscopy images of as-prepared CNF film (PAN-220 + 550). Reproduced from Ref. [7]. Copyright 2013, WILLEY-VCH

1.3.2   Aligned Fibers

Due to the several merits (example: high strength) of aligned fibers, is applied in various applications. To harvest aligned fibers, quite a lot of new set-ups are compulsory, specifically on collectors. Commonly, drum collector with high speed used to yield aligned fibers. For example, Matthews et al. initially witnessed the alignment of collagen fibers at a rotation speed of 1.4 m/s. However, it is hard to get impeccably aligned fibers. Therefore, another approach reported practicing two stripes of parallel conductors as collectors. Interestingly, Li et al. prepared uniaxial alignment nanofibers over a gap formed between two conductive substrates and the images shown in Fig. 1.4.

Fig. 1.4 Dark-field optical micrograph of PVP nanofibers collected on top of a gap formed between two silicon stripes. Reproduced from Ref. [41]. Copyright 2003, American Chemical Society

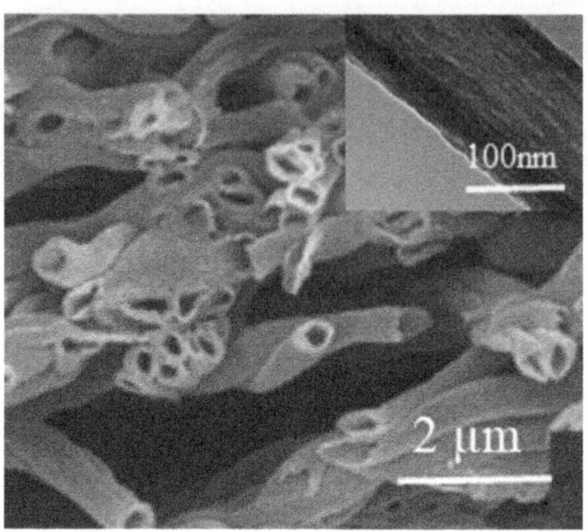

Fig. 1.5 The cross-sectional SEM and TEM images for Meso-HACNF2 core–shell fibers. Reproduced from Ref. [44]. Copyright 2013, Elsevier B.V.

1.3.3   Core/Shell Fibers

Regard to core shell structured fibers; it has drawn quick interest in various applications, for example, electrochemical energy storage applications. Broadly, the core acts as inner functional materials, whereas shell assists as a shielding layer [42]. It is recommended that the electrospinning engaging with dual nozzles offers a superficial mode to synthesis unremitting, core/shell structured fibers on a massive scale. Previously, two different polymer solutions have been used using a co-axial nozzle [43]. For example, Park et al. proposed activated carbon nanofibers with hollow core/highly mesoporous shell structure in which poly (methyl methacrylate) (PMMA) as a core precursor and with either polyacrylonitrile (PAN) or PAN/PMMA blended polymer as a shell precursor subsequent stabilization, carbonization, and activation process. As prepared mesopores of activated carbon nanofibers with hollow core/porous shell structure are shown in Fig. 1.5, that has a core approximately 200 nm and the shell has a 360 nm in diameter, respectively [44].

1.4   Applications

Inimitable characteristics of chemical and physical properties of carbon nanofibers qualifying their claims in these three significant fields such as tissue engineering, water treatment and energy storage applications. In brief to tissue engineering, the carbon nanofibers are of excessive importance in regenerative medicine on account of their structural resemblance to that found in a native extracellular matrix (ECM).

What is more, the interlocked pores and high surface area to volume ratios could deliver extra room for the adoption of cells, and sustain their activities like migration, proliferation, and differentiation. The compelling communications between cells and the nearby microenvironments occur all the time through tissue homeostasis. Consequently, accepting the biophysical and biochemical relations among cells and the ECM is essential to managing cellular activities for capable tissue regeneration [9]. Hence, information of electrospun nanofibers hallmarks in terms of surface chemistry, fiber tomography, mechanical, and physical properties will offer a consistent model of fine-tuning the nanofibers for usage as appropriate implants like skin graft [45], nerve graft [46] cardiac cells graft [47] and bone graft [48] regeneration.

Taking account on water treatment, at this moment growing request and scarcity of clean water bases because of quick industrialization, population growth, and determined famine environments have become worldwide severe matters. About the environment is worried, the recycling of wastewater is required to boost the inadequate freshwater stock and to encounter the new strain on water resources in the long run. Over the past few decades, so many approaches have been suggested to improve the practicable wastewater treatment process [49, 50]. Besides, as grown in Fig. 1.6, carbon nanofibers have been established as auspicious aspirants for heavy-ion adsorbents and photocatalysis of organic pollutants in water treatment, by virtue of their

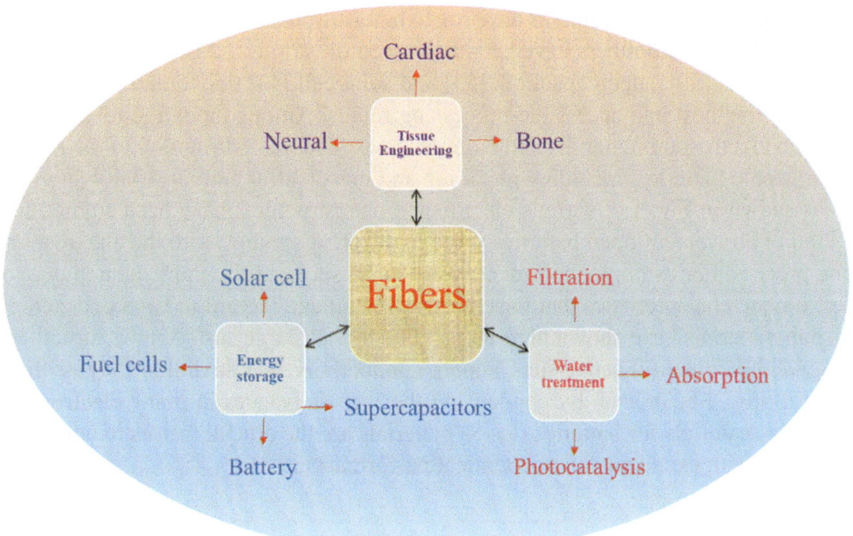

Fig. 1.6 Elucidation of electrospun carbon nanofibers and their hybrid composites for the various applications

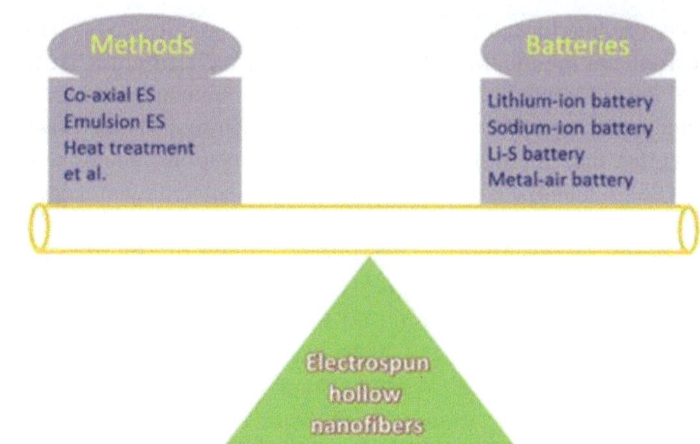

Fig. 1.7 A schematic diagram of electrospun hollow nanofibers for advanced secondary batteries. Reproduced from Ref. [1]. Copyright 2013, Elsevier B.V.

high porosity, interconnectivity, and a large surface-to-volume ratio [51]. Perspectives on energy storage, its electrochemical performance depends on the electrode materials. During the past decades, materials science has received considerable devotion, and the rapid improvements have run to fantastic progress in energy storage. For example, carbon nanofibers have been addressed in various fields such as fuel cells [52], batteries [20], supercapacitors [53] and solar cell [44] due to the 1D structure and their low cost and accessible processing routes. Among them, the relationship between electrospun materials and batteries reveals more contribution to the recent rechargeable batteries, including metal ion and metal sulfur and metal-air batteries.

As shown in Fig. 1.7, various electrospun nanostructures have been studied for al kind of abovementioned batteries through different strategy with the electrospinning process. For example, hollow nanostructures successfully implemented due to their unique characteristics that have been systematical. Regard to Li–S cells; electrospun materials have shown high impact due to their large surface area, structural designability, plentiful active sites, superb conductivity, and flexibility [54] as illustrated in Fig. 1.8. Indeed, by condensing the latest advances in many electrospun nanofiber materials for batteries (Li–S) materials are the crucial factors to construct Li–S cells with excellent electrochemical performance.

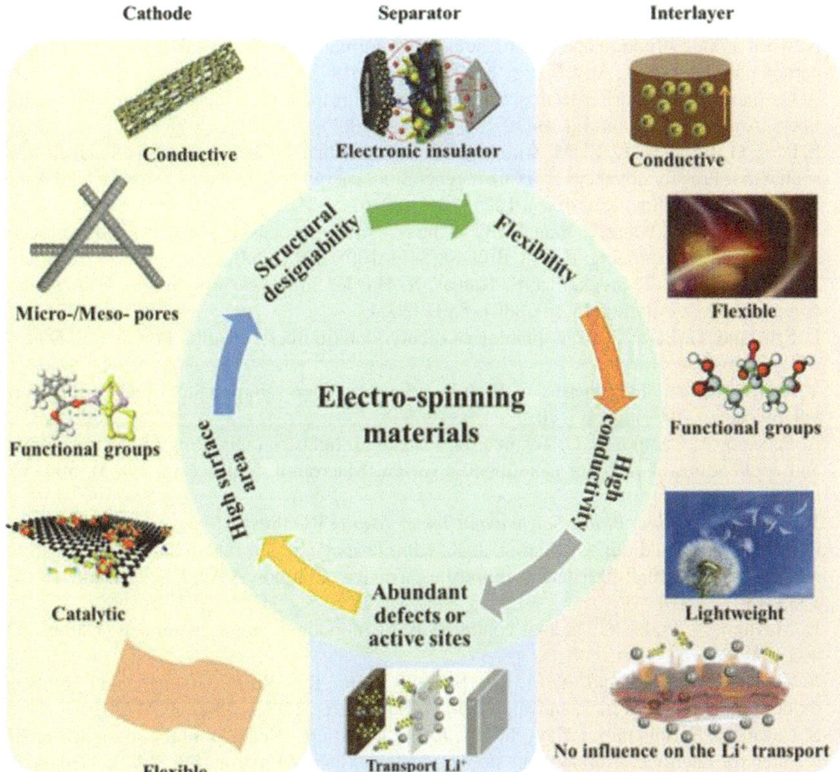

Fig. 1.8 Unique properties of electrospun nanofiber materials for Li–S batteries. Reproduced from Ref. [54]. Copyright 2013, Elsevier B.V.

References

1. L. Li, S. Peng, J.K.Y. Lee, D. Ji, M. Srinivasan, S. Ramakrishna, Electrospun hollow nanofibers for advanced secondary batteries. Nano Energy **39**, 111–139 (2017)
2. M. Winter, R.J. Brodd, *What are Batteries, Fuel Cells, and Supercapacitors?* (ACS Publications, 2004)
3. D.S. Su, R. Schlögl, Nanostructured carbon and carbon nanocomposites for electrochemical energy storage applications, ChemSusChem: Chem. Sustain. Energy Mater. **3**(2), 136–168 (2010)
4. B. Zhang, Y. Yu, Y. Liu, Z.-D. Huang, Y.-B. He, J.-K. Kim, Percolation threshold of graphene nanosheets as conductive additives in Li 4 Ti 5 O 12 anodes of Li-ion batteries. Nanoscale **5**(5), 2100–2106 (2013)
5. X. Li, F. Kang, X. Bai, W. Shen, A novel network composite cathode of LiFePO4/multiwalled carbon nanotubes with high rate capability for lithium ion batteries. Electrochem. Commun. **9**(4), 663–666 (2007)
6. I. Martin-Gullon, J. Vera, J.A. Conesa, J.L. González, C. Merino, Differences between carbon nanofibers produced using Fe and Ni catalysts in a floating catalyst reactor. Carbon **44**(8), 1572–1580 (2006)

7. B. Zhang, Y. Yu, Z.L. Xu, S. Abouali, M. Akbari, Y.B. He, F. Kang, J.K. Kim, Correlation between atomic structure and electrochemical performance of anodes made from electrospun carbon nanofiber films. Adv. Energy Mater. **4**(7), 1301448 (2014)

8. A. Greiner, J.H. Wendorff, Electrospinning: a fascinating method for the preparation of ultrathin fibers. Angew. Chem. Int. Ed. **46**(30), 5670–5703 (2007)

9. S. Peng, G. Jin, L. Li, K. Li, M. Srinivasan, S. Ramakrishna, J. Chen, Multi-functional electro-spun nanofibres for advances in tissue regeneration, energy conversion and storage, and water treatment. Chem. Soc. Rev. **45**(5), 1225–1241 (2016)

10. X.-G. Dong, C.-G. Wang, C. Juan, Study on the coagulation process of polyacrylonitrile nascent fibers during wet-spinning. Polym. Bull. **58**(5–6), 1005–1012 (2007)

11. X. Zhang, T. Liu, T. Sreekumar, S. Kumar, X. Hu, K. Smith, Gel spinning of PVA/SWNT composite fiber. Polymer **45**(26), 8801–8807 (2004)

12. L. Szosland, G. East, The dry spinning of dibutyrylchitin fibers. J. Appl. Polym. Sci. **58**(13), 2459–2466 (1995)

13. Y. Arai, Structure and properties of pitch-based carbon fibers. Nippon Steel Technical Report, vol. 59 (Japan, 1993), pp. 65–70

14. L. Persano, A. Camposeo, C. Tekmen, D. Pisignano, Industrial upscaling of electrospinning and applications of polymer nanofibers: a review. Macromol. Mater. Eng. **298**(5), 504–520 (2013)

15. J.-B. Donnet, *Carbon Black: Science and Technology* (CRC Press, 1993)

16. P.-C. Ma, N.A. Siddiqui, G. Marom, J.-K. Kim, Dispersion and functionalization of carbon nanotubes for polymer-based nanocomposites: a review. Compos. A Appl. Sci. Manuf. **41**(10), 1345–1367 (2010)

17. E. Mayhew, V. Prakash, Thermal conductivity of individual carbon nanofibers. Carbon **62**, 493–500 (2013)

18. Y. Zhu, S. Murali, W. Cai, X. Li, J.W. Suk, J.R. Potts, R.S. Ruoff, Graphene and graphene oxide: synthesis, properties, and applications. Adv. Mater. **22**(35), 3906–3924 (2010)

19. S. Cavaliere, S. Subianto, I. Savych, D.J. Jones, J. Rozière, Electrospinning: designed archi-tectures for energy conversion and storage devices. Energy Environ. Sci. **4**(12), 4761–4785 (2011)

20. B. Zhang, F. Kang, J.-M. Tarascon, J.-K. Kim, Recent advances in electrospun carbon nanofibers and their application in electrochemical energy storage. Prog. Mater Sci. **76**, 319–380 (2016)

21. S. Peng, L. Li, J.K.Y. Lee, L. Tian, M. Srinivasan, S. Adams, S. Ramakrishna, Electrospun carbon nanofibers and their hybrid composites as advanced materials for energy conversion and storage. Nano Energy **22**, 361–395 (2016)

22. M. Lallave, J. Bedia, R. Ruiz-Rosas, J. Rodríguez-Mirasol, T. Cordero, J. Otero, M. Marquez, A. Barrero, I.G. Loscertales, Adv. Mater. **19**, 4292–4296 (2007)

23. J.A. Matthews, G.E. Wnek, D.G. Simpson, G.L. Bowlin, Electrospinning of collagen nanofibers. Biomacromolecules **3**(2), 232–238 (2002)

24. D. Zhu, C. Xu, N. Nakura, M. Matsuo, Study of carbon films from PAN/VGCF composites by gelation/crystallization from solution. Carbon **40**(3), 363–373 (2002)

25. J.R. Kim, S.W. Choi, S.M. Jo, W.S. Lee, B.C. Kim, Electrospun PVdF-based fibrous polymer electrolytes for lithium ion polymer batteries. Electrochim. Acta **50**(1), 69–75 (2004)

26. U.K. Fatema, A.J. Uddin, K. Uemura, Y. Gotoh, Fabrication of carbon fibers from electrospun poly (vinyl alcohol) nanofibers. Text. Res. J. **81**(7), 659–672 (2011)

27. W.N.W. Salleh, A.F. Ismail, Fabrication and characterization of PEI/PVP-based carbon hollow fiber membranes for CO2/CH4 and CO2/N2 separation. AIChE J. **58**(10), 3167–3175 (2012)

28. Y.-E. Miao, G.-N. Zhu, H. Hou, Y.-Y. Xia, T. Liu, Electrospun polyimide nanofiber-based nonwoven separators for lithium-ion batteries. J. Power Sources **226**, 82–86 (2013)

29. X. Zhang, P. Suresh Kumar, V. Aravindan, H.H. Liu, J. Sundaramurthy, S.G. Mhaisalkar, H.M. Duong, S. Ramakrishna, S. Madhavi, Electrospun TiO2–graphene composite nanofibers as a highly durable insertion anode for lithium ion batteries. J. Phys. Chem. C **116**(28), 14780–14788 (2012)

30. Y. Yu, L. Gu, C. Zhu, P.A. Van Aken, J. Maier, Tin nanoparticles encapsulated in porous multichannel carbon microtubes: preparation by single-nozzle electrospinning and application as anode material for high-performance Li-based batteries. J. Am. Chem. Soc. **131**(44), 15984–15985 (2009)
31. S.H. Park, C. Kim, Y.O. Choi, K.S. Yang, Preparations of pitch-based CF/ACF webs by electrospinning. Carbon **13**(41), 2655–2657 (2003)
32. X.F. Li, Q. Xu, Y. Fu, Q.X. Guo, Preparation and characterization of activated carbon from Kraft lignin via KOH activation. Environ. Prog. Sustain. Energy **33**(2), 519–526 (2014)
33. L. Zou, L. Gan, F. Kang, M. Wang, W. Shen, Z. Huang, Sn/C non-woven film prepared by electrospinning as anode materials for lithium ion batteries. J. Power Sources **195**(4), 1216–1220 (2010)
34. R. Ruiz-Rosas, J. Bedia, M. Lallave, I. Loscertales, A. Barrero, J. Rodríguez-Mirasol, T. Cordero, The production of submicron diameter carbon fibers by the electrospinning of lignin. Carbon **48**(3), 696–705 (2010)
35. P. Gupta, C. Elkins, T.E. Long, G.L. Wilkes, Electrospinning of linear homopolymers of poly (methyl methacrylate): exploring relationships between fiber formation, viscosity, molecular weight and concentration in a good solvent. Polymer **46**(13), 4799–4810 (2005)
36. C. Wang, C.-H. Hsu, J.-H. Lin, Scaling laws in electrospinning of polystyrene solutions. Macromolecules **39**(22), 7662–7672 (2006)
37. H. Fong, I. Chun, D.H. Reneker, Beaded nanofibers formed during electrospinning. Polymer **40**(16), 4585–4592 (1999)
38. X. Li, Y. Chen, H. Huang, Y.-W. Mai, L. Zhou, Electrospun carbon-based nanostructured electrodes for advanced energy storage–a review. Energy Storage Mater. **5**, 58–92 (2016)
39. D. Li, Y. Xia, Electrospinning of nanofibers: reinventing the wheel? Adv. Mater. **16**(14), 1151–1170 (2004)
40. W.E. Teo, S. Ramakrishna, A review on electrospinning design and nanofibre assemblies. Nanotechnology **17**(14), R89 (2006)
41. D. Li, Y. Wang, Y. Xia, Electrospinning of polymeric and ceramic nanofibers as uniaxially aligned arrays. Nano Lett. **3**(8), 1167–1171 (2003)
42. D. Li, A. Babel, S.A. Jenekhe, Y. Xia, Nanofibers of conjugated polymers prepared by electrospinning with a two-capillary spinneret. Adv. Mater. **16**(22), 2062–2066 (2004)
43. Z. Sun, E. Zussman, A.L. Yarin, J.H. Wendorff, A. Greiner, Compound core–shell polymer nanofibers by co-electrospinning. Adv. Mater. **15**(22), 1929–1932 (2003)
44. S.-H. Park, B.-K. Kim, W.-J. Lee, Electrospun activated carbon nanofibers with hollow core/highly mesoporous shell structure as counter electrodes for dye-sensitized solar cells. J. Power Sources **239**, 122–127 (2013)
45. F. Groeber, M. Holeiter, M. Hampel, S. Hinderer, K. Schenke-Layland, Skin tissue engineering—in vivo and in vitro applications. Adv. Drug Deliv. Rev. **63**(4–5), 352–366 (2011)
46. S.H. Lim, X.Y. Liu, H. Song, K.J. Yarema, H.-Q. Mao, The effect of nanofiber-guided cell alignment on the preferential differentiation of neural stem cells. Biomaterials **31**(34), 9031–9039 (2010)
47. S. Mukherjee, J. Reddy Venugopal, R. Ravichandran, S. Ramakrishna, M. Raghunath, Evaluation of the biocompatibility of PLACL/collagen nanostructured matrices with cardiomyocytes as a model for the regeneration of infarcted myocardium. Adv. Funct. Mater. **21**(12), 2291–2300 (2011)
48. W. Liu, J. Lipner, C.H. Moran, L. Feng, X. Li, S. Thomopoulos, Y. Xia, Generation of electrospun nanofibers with controllable degrees of crimping through a simple. Plasticizer-Based Treat. Adv. Mater. **27**(16), 2583–2588 (2015)
49. M.M. Khin, A.S. Nair, V.J. Babu, R. Murugan, S. Ramakrishna, A review on nanomaterials for environmental remediation. Energy Environ. Sci. **5**(8), 8075–8109 (2012)
50. A. Sánchez, S. Recillas, X. Font, E. Casals, E. González, V. Puntes, Ecotoxicity of, and remediation with, engineered inorganic nanoparticles in the environment. TrAC, Trends Anal. Chem. **30**(3), 507–516 (2011)

51. J. Deng, X. Kang, L. Chen, Y. Wang, Z. Gu, Z. Lu, A nanofiber functionalized with dithizone by co-electrospinning for lead (II) adsorption from aqueous media. J. Hazard. Mater. **196**, 187–193 (2011)
52. D.S. Yang, S. Chaudhari, K.P. Rajesh, J.S. Yu, Preparation of nitrogen-doped porous carbon nanofibers and the effect of porosity, electrical conductivity, and nitrogen content on their oxygen reduction performance. ChemCatChem **6**(5), 1236–1244 (2014)
53. C. Kim, Electrochemical characterization of electrospun activated carbon nanofibres as an electrode in supercapacitors. J. Power Sources **142**(1–2), 382–388 (2005)
54. M. Liu, N. Deng, J. Ju, L. Fan, L. Wang, Z. Li, H. Zhao, G. Yang, W. Kang, J. Yan, A review: electrospun nanofiber materials for lithium-sulfur batteries. Adv. Funct. Mater. (2019)

Chapter 2
Electrospinning of Nanofibers for Li-Ion Battery

Abstract Li-ion batteries have been widely used in portable electronic devices owing to their excellent assets such as high energy/power density and long operating life. There is an urgent demand to develop high-performance electrode materials with the high capability to store and distribute energy more effectively to additional boost the performance of LIBs. Up to now, numerous electrode materials have been cast-off to achieve high performance. Among them, electrospun based anode and cathode materials have a great interest in both energy and power demands of rechargeable LIBs. In this chapter, we focused the recent research advances in the electrospun related anode and cathode materials for the next generation of LIBs. Remarkably, we highlighted the electrochemical properties of electrodes include metal, metal oxides, phosphides, sulfides, nitrides for LIBs applications. Besides, the unique features like a high surface area, particle size low diffusion distance, high electrical, ionic conductivity, cycling stability were discussed. Also, and binder-free electrodes fabrication method formation of porous network and growth for the flexible LIBs were studied. Based on their electrochemical reaction with lithium, intercalation/de-intercalation, alloy/de-alloy, and conversion materials with a wide range of recent research progress and perspectives on anode materials discussed in more detail.

Keywords Binder-free electrode · Surface morphology · Li-ion battery

2.1 Li-Ion Battery Working Principle and Cell Structure

Rechargeable Li-ion batteries (LIBs) working standard can be enlightened with a schematic diagram shown in Fig. 2.1. The working principle is based on the 'host/guest' reaction mechanism. The Li-ion (Li^+) shuttles between cathodes to the anode while charge (Oxidation and de-lithiation), concurrently, the movement of the electron via an external circuit to reduce the anode and this process is reversible during the discharge. LIBs composed of an anode (main graphite), a cathode (Lithium metal oxide), Electrolyte (Non-aqueous generally contains $LiFP_6$ salt in ethylene carbonate and dimethyl carbonate solution) and Separator (normally Polypropylene).

S. Peng and P. R. Ilango, *Electrospinning of Nanofibers for Battery Applications*, https://doi.org/10.1007/978-981-15-1428-9_2

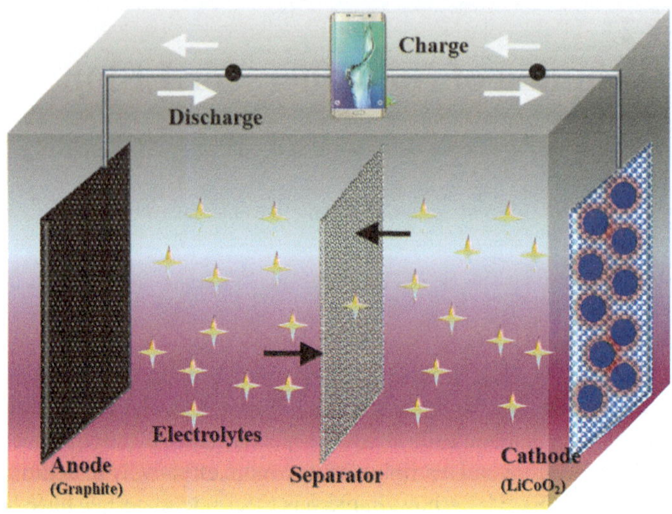

Fig. 2.1 Schematic diagram of typical rechargeable Li-ion battery

In the past decades, LIBs dominating power sources to practical applications in portable electronics and HEVs due to their high energy and power density as well as it has correct cell dimensions. The cathode and anode determine power and energy density of the system. At the same time, the non-aqueous electrolyte is responsible for offering paths to Li^+ transport, and finally, the separator prevents from short circuit. The Li metal is the most electropositive (-3.04 vs. standard hydrogen electrode) and lightweight material (equivalent to $M = 6.94$ g mol^{-1}) [1]. The $Li_{1-x}CoO_2$ was introduced as a good intercalated cathode and intercalated carbon as anode material in the 1980s and commercialized 1991s by Sony [2]. The electrode reaction at discharge can be expressed as [3]

$$Li_x C_6 \rightarrow x Li^+ + 6C + xe^-$$

(2.1)

$$Li_{1-x}Co_2 + xLi^+ + xe^- \rightarrow LiCoO_2$$

(2.2)

There are many stages involving during the intercalation of Li into graphite such as LiC_{24}, LiC_{27} and LiC_{12} via first-order phase transaction reactions [4]. While, from the electrode reaction x is intentionally defined to 0.5 because of safety issues, related to 4.2 V versus Li/Li$^+$ [1]. The working mechanism of LIBs accepts a thermodynamic law. Both densities are associated with LIBs output voltage, deliverable current, and time. The output voltage would be measured by electrochemical potentials of electrode materials (i.e. anode and cathode) expressed as [5]

$$V_{max} = \frac{\mu_a - \mu_c}{e}$$

(2.3)

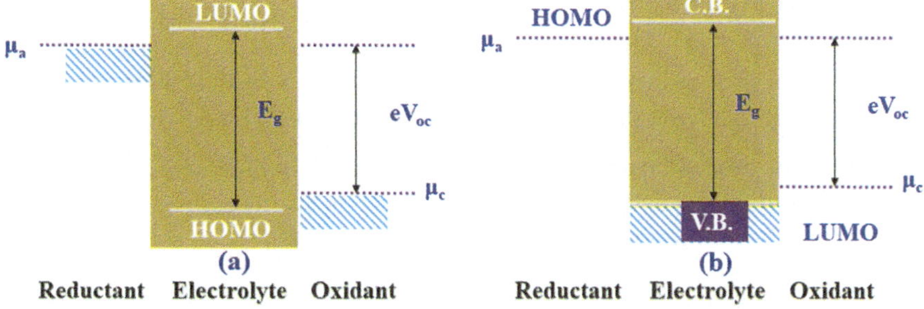

Fig. 2.2 The energy gap between LUMO and HOMO. **a** Liquid electrolyte with solid electrodes; **b** solid electrolyte with liquid or gaseous reactants. Reproduced from Ref. [3]. Copyright 2013, American Chemical Society

where μ_a and μ_b are the anode and cathode electrochemical potential and 'e' magnitude of the electron charge. The electrochemical potential involved in both energy and Li-ion transfer. The output voltage is by the "window" of the electrolyte. The window of the electrolyte is depicted in Fig. 2.2a, b, is the energy gap between the lowest unoccupied and highest occupied molecular orbitals (LUMO and HOMO) of a liquid electrolyte. As represented in Fig. 2.2a, b, a μ_a above the electrolyte LUMO reduces the electrolyte unless the anode-electrolyte reaction becomes jammed by the formation of a passivating SEI layer; likewise, a 'μ_c' situated underneath the HOMO oxidizes the electrolyte unless the reaction is jammed by an SEI layer. Conversely, μ_c cannot be dropped below the top of the cathode anion-p bands, which may have an energy above the electrolyte HOMO. The top of the sulfide S-3p bands of the layered sulfides $LiMS_2$ is $\sim$2.5 eV below the μa of a Li anode, μa (Li), whereas the top of the O-2p bands of the layered oxides $LiMO_2$ is $\sim$4.0 eV below μ_a(Li), which is why oxide hosts are used as positive electrodes of contemporary day Li-ion batteries [6, 7] Since the practical HOMO of the organic liquid carbonate electrolytes used in Li-ion batteries is at 4.3 eV below μa(Li), the voltage of the simple $LiMO_2$ layered oxides is also self-limited by the energy of the top of the O-2p bands [3].

2.2 Perspectives on Material Development

2.2.1 Cathode

In recent years, rechargeable LIB systems have become a leading technology in the world-wide battery market. These batteries supply the maximum energy density obtainable to date for secondary batteries. But ever-increasing the usage of LIBs push to demand large-scale applications. The foremost aspect that decides the energy density, rate capability (i.e., power density), and price of LIBs is the cathode [8]. Varieties of positive materials have been developed and are available commercially such as $LiCoO_2$ [9], $LiMn_2O_4$ [10], $LiFePO_4$ [11], $LiMn_{1-x-y}Ni_xCo_yO_2$ [12], Li_2MSiO_4 [13] and $LiMBO_3$ [14]. The intercalation cathode materials are mainly classified into three types, such as layered, spinel, and olivine by their crystal lattice types. The famous cathode materials should have the following features [15]

(i)   High redox potential and high Li-ions storing to deliver enough capacity.
(ii) Fast insertion and extraction of Li-ions and high electronic conductivity
(iii) Cheap, environmentally benign and ease to synthesis.

Thus, materials optimizations are usually made from two main features, to change the intrinsic chemistry and to change the morphology of the materials. Figure 2.3, associates the gravimetric energy densities of different cathode materials that are now under research. Meanwhile, some materials such as $LiFeBO_3$ and $LiFeSO_4F$ are already imminent their theoretical energy densities, for other materials comprising conventional layered and spinel compounds, noteworthy gaps are still existing between their theoretical and practical energy densities. The materials with encouraging theoretical properties have high potentials as the candidates of future generation LIB cathode, therefore are under intensive studies [16].

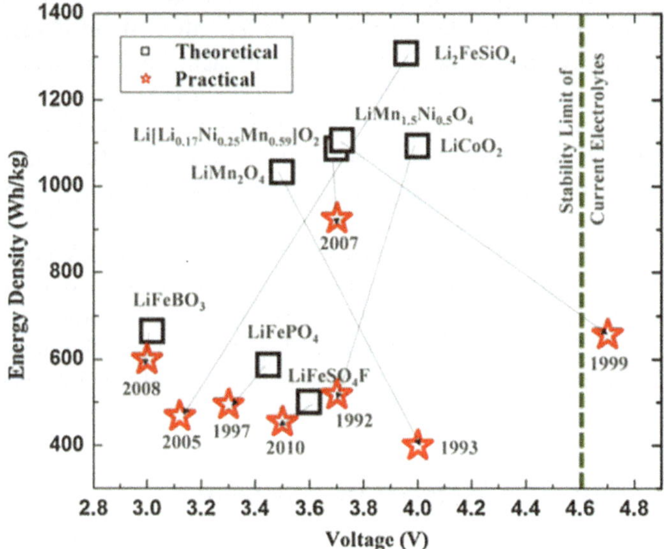

Fig. 2.3 Theoretical and practical gravimetric energy densities of different cathode materials. Reproduced from Ref. [16]. Copyright 2012, Elsevier

2.2.2 Anode

For reversible intercalation of Li^+, commercial rechargeable batteries depend on anodes prepared by graphite, essentially in the form of particles (approximately 15–20) μm. Even though the decent performances of the graphite in respect of electronic conductivity, low electrochemical potential, and coulombic efficiency (>95%). Moreover, graphite possesses a low theoretical capacity (372 mAh/g) [17]. Regrettably, it cannot meet the up-to-date increasing demand challenges. Therefore, many research is now focused on new anode materials such as Si, Sn, Sb and Ge [18, 19]. Anode materials have been classified as follows,

$$xLi^+ + C_6(\text{in graphite}) + xe^- \leftrightarrow Li_xC_6(\text{intercalated anode}) \qquad (2.4)$$

$$M_aX_b + (b.n)\,Li^+ + ae^- \leftrightarrow aM + bLi_nX \text{ (conversion anodes)} \qquad (2.5)$$

$$M + xLi^+ + xe^- \leftrightarrow Li_xM \text{ (alloy anodes)} \qquad (2.6)$$

Excitingly, the conversion type anodes (mostly transition metals) and alloy anodes deliver high storage capacity than commercial graphite. For instance, $Li_{4.4}Si$ is 4200 mAh/g, $Li_{4.4}Ge$ is 1600 mAh/g, $Li_{4.4}Sn$ is 990 mAh/g, and Li_3Sb is 660 mAh/g corresponding to Li storage, respectively [20]. Nonetheless, their efficiencies of discharge to charge are not as decent as that of graphite [21]. The origin of poor efficiency associated because of two main aspects (i) huge volume change (Example Si is 320%) during the intercalation and de-intercalation process that causes the fracturing, pulverization, and poor electrical contact (ii) unremitting formation and reformation of the solid electrolyte interfaces (SEI) during the electrochemical process [22]. This occurrence will seriously limit their potential application such as commercial electronic items. Several strategies have been recommended in the last years to reduce the confines due to both the volume swelling under Li charging and the intrinsic low electrical conductivity. Hence, the track foremost to LIBs with better-quality energy and power density has, as a major task, the assortment of appropriate anode materials which can deliver high capacity and comfort diffusion of Li^+ into the anode, along with enhanced calendar life [23]. Figure 2.4 represents the potential and capacity of different anode active materials for next-generation LIB anodes.

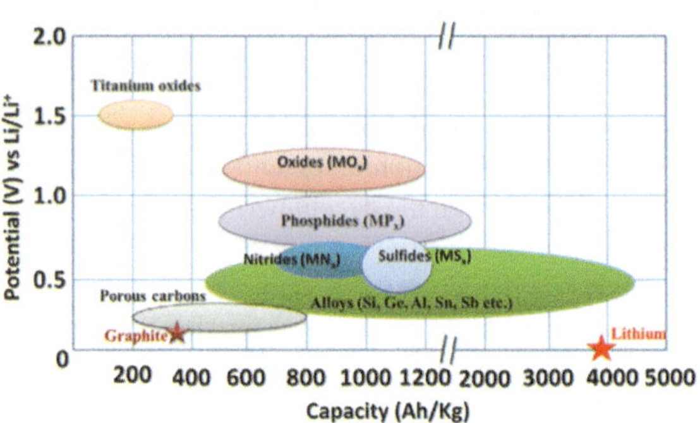

Fig. 2.4 Schematic representation of active anode materials for the next generation of LIBs. Reproduced from Ref. [23]. Copyright 2013, Elsevier

2.3 Electrodes

2.3.1 Cathodes

Exploiting new cathode materials with high specific capacities is of excellent repu-
tation in the development of flexible LIBs. Vanadium based materials are broadly
examined cathode materials for LIBs, but these materials typically suffer from imper-
fect electrochemical performances and still cannot adequately fulfil the increasing
difficulties [24, 25]. The poor kinetic properties of LVP caused by its intrinsic low
electronic conductivity (2×10^{-8} S cm^{-1}) and ionic conductivity (10^{-9} to 10^{-10} cm^2
s^{-1}) hinder its commercial usage [26, 27]. Several approaches have been developed
to relieve this issue. However, those methods presented inadequate success, but it
remains an excellent task to seek effective strategies. For instance, *P*. Sun et al.
fabricated C-LVP/CNFs in which LVP nanoparticle encapsulated carbon nanofibers.
What's more, plasma enhanced chemical vapor deposition (PECVD) of acetylene
(C_2H_2) was used to add a smooth thin layer of carbon to improve further the conductivity of carbon coated LVP/CNFs. the thickness of the carbon layer was estimated to
be roughly 6–7 nm which is observed from the SEM images shown in Fig. 2.5a. The
C-LVP/CNF cathodes revealed a minimal capacity decrease at higher rates, demonstrating their excellent high-rate capability. Even at 10 C, a discharge capacity of
120 mAh g^{-1} (90.5% of the theoretical capacity) was delivered by the C-LVP/CNFs,
almost two times higher than that of the LVP/CNFs (64 mAh g^{-1}) demonstrated in
Fig. 2.5b which proves that a conductive carbon outer-layer of composite cathodes
can significantly enhance the rate capability and long-term cyclability [28]. Also,
nickel sulfide (NiS) is one of the most capable cathode materials owing to its excellent stability over a wide range of temperature and a high theoretical capacity of
590 mAh g^{-1}. Li et al. reported a feasible strategy to synthesize hybrid materials
by in situ polymerizations of a conducting polymer of polypyrrole (PPy) on NiS-
carbon nanofiber films to form high-performance LIB cathodes for the first time.
TEM study indicates that the thickness of the conformal PPy coating layer on the
surface of the NiS particle is around 20 nm illustrated in Fig. 2.5c. The potential
profiles of the NiS-PPy-CNF hybrid electrode depicted in Fig. 2.5d with a three-step

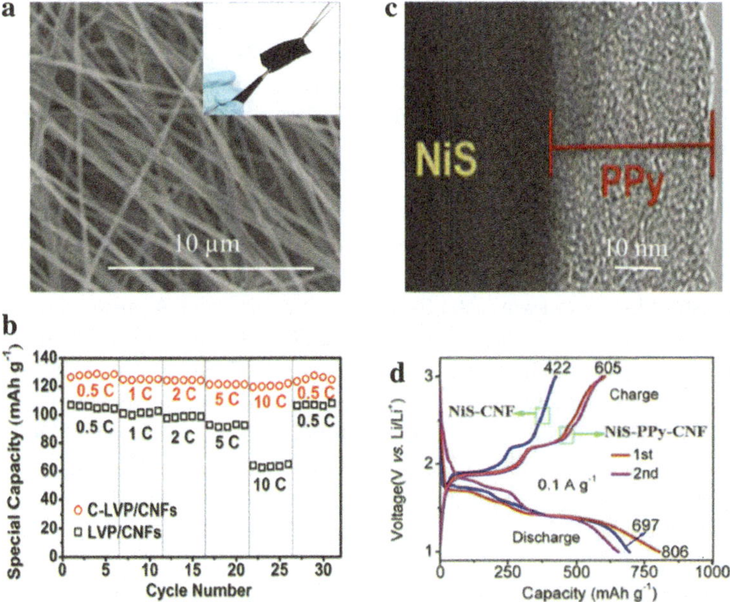

Fig. 2.5 **a** SEM images of C-LVP/CNFs. **b** Rate performance of C-LVP/ CNFs and LVP/CNFs. Reproduced from Ref. [28]. Copyright 2016, The Royal Society of Chemistry. **c** HRTEM images showing crystalline NiS particles coated by a conformal PPy layer. **d** The charge/discharge profiles of NiS-CNF and NiS-PPy-CNF. Reproduced from Ref. [29]. Copyright 2016, Elesvier

discharging process at around 1.8, 1.7, and 1.4 V, respectively. The first two peaks at higher potential correspond to the transformation of NiS to Ni_3S_2 ($3NiS + 2Li \leftrightarrow Ni_3S_2 + Li_2S$) and the last peak at lower potential can be attributed to the conversion of Ni_3S_2 to Ni ($Ni_3S_2 + 4Li \leftrightarrow 3Ni + 2Li_2S$). The charge process spectacles two charge plateaus located at ~1.9 and 2.2 V, which are attributed to the regeneration of NiS. Such electrochemical actions are in good treaty with the results of cyclic voltammograms tests [29].

2.3.2 Free-Standing Anodes

Silicon (Si) is one of the promising anode material to replace commercial graphite in rechargeable Li-ion batteries owing to its high theoretical capacity of 4200 mAh g^{-1} corresponding to $Li_{4.4}Si$ and rich resources [30, 31]. Nevertheless, some challenges for the Si anode are also apparent due to the low conductivity and large volume expansion (about 300%) during the cycling. This enlargement difficult origins the pulverization of Si and the frequent formation of the solid electrolyte interface (SEI) on Si anode, ensuing in the loss of electrical contact between the inter particles and subsequently fading the battery life cycle [30]. To overcome volume expansion of Si

Si-based materials, extensive research had been carried out. Nanosized Si materials allow hopeful chance to deal with all these challenges because of their potential to relax strain. Moreover, the synthesis of various morphologies such as nanowires, nanoflower, porous structure and composites have been demonstrated to resolve these problems [32]. Developing novel free-standing electrodes via electrospinning process illustrates the fast transfer of Li$^+$ and electrons. Also, electrodes commendably delayed re-aggregation of metal nanoparticles during lithiation/delithiation. For instance, Kong et al. prepared Si nanoparticles encapsulated within the network of thin graphitized carbon layer (from carbonized polydopamine, named as C-PDA) through a straightforward way, establishing free-standing mat. As an anode, it delivers a high capacity of 1750 mAh g^{-1} after 100 cycles at 100 mA g^{-1}, respectively [31]. Similarly, Xu et al. fabricated free-standing Si/Ni/CNF electrodes containing ultra-fine Ni conductive particles through the same electrospinning process. The coating of the active Si particles sustained the stability of SEI layers, which subsidized to the exceptional cyclic like of the Si/Ni/CNF composites with notable specific capacity of 1045 mAh g^{-1} at the 50th cycle at a current density of 100 mA g^{-1} [33]. Later, the same research group had published free-standing Si/CNF/G composite without adding of Ni element in which Si nanoparticles are well bonded to the polymer precursor and fully encapsulated within amorphous carbon after carbonization method. Figure 2.6a depicts the SEM images of the as-prepared 5PVA-1TSi-GO sample, which has 1% of extra G.O. sheets added does not affect the morphology. The corresponding cycling performance reveals that the TSi/CNF/G composite electrodes

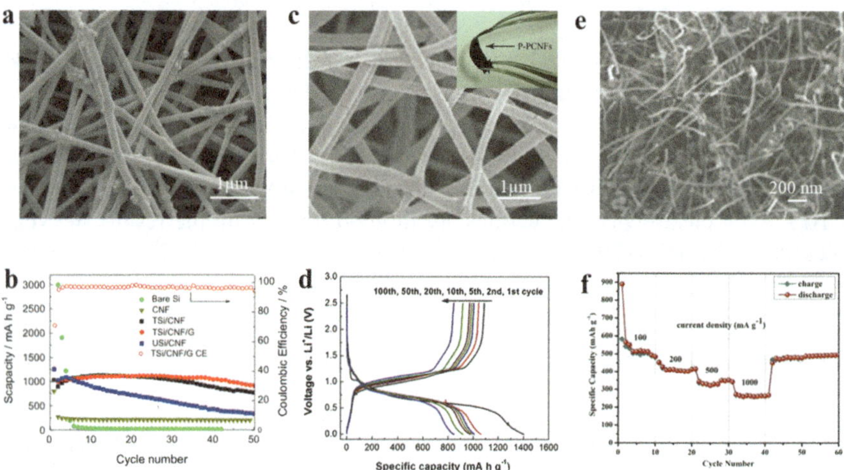

Fig. 2.6 **a** SEM images of 5PVA–1TSi–G.O. **b** Cyclic performance of electrodes made from neat Si, CNF, USi/CNF, TSi/CNF and TSi/CNF/G at a current density of 100 mA g^{-1}. Reproduced from Ref. [34]. Copyright 2014, Elsevier. **c** The FE-SEM micrographs of P-PCNFs. **d** Voltage profiles of P-PCNFs. Reproduced from Ref. [36]. Copyright 2014, Elsevier. **e** SEM images of CNTs grown on CNFs surface (30 s). **f** Rate performance of Ni-CNFs-CNTs. Reproduced from Ref. [40]. Copyright 2019, Elsevier

delivered a significant reversible capacity above 872 mAh g^{-1} after 50 cycles with an unpaid retention rate of 91% shown in Fig. 2.6b [34]. Recently, Si/Carbon nanofibers composite membrane with high mechanical strength is produced by electrospinning and chemical vapor deposition techniques. This three-dimensional network membrane has acted a vital role in alleviating volume expansion–contraction of Si-based anodes [35]. Elemental phosphorus (P) silicon has quite small atomic weights, and higher theoretical specific capacity of 2595 mAh g^{-1} for Li_3P and it has been used flexible and free-standing for LIBs. Initially, Li et al. reported that P-PCNFs in which porous carbon nanofibers encapsulated with crystalline red P as a anode. The FE-SEM images and photographs of P-PCNFs are shown in Fig. 2.6c (inset). It displays a uniform diameter of 200 nm with interconnected structure and Fig. 2.6d illustrates the discharge–charge voltage profiles. It can be noted that the first discharge and charge cycle for P-PCNFs composite electrode is 1402 and 1088 mAh g^{-1}, respectively, giving an initial coulombic efficiency of 78% which is mainly ascribed to the irreversible formation of SEI layer. Furthermore, the capacity loss only chances in the first several cycles; after that, the C.E. approaches 100%, indicating the superb capacity retention of P-PCNFs [36]. Afterwards, Li et al. discovered flexible P-doped carbon nanosheets/nanofibers as paper using black phosphorus (B.P.) as a doping source via electrospinning and subsequently an evaporation/deposition process. After the electrochemical investigations, they reported that the optimal sample BPCNF-900 shows a higher reversible capacity of 1100 at 200 mA g^{-1} after 550 cycles [37]. Recently, Red P has been confined within the porous carbon derived from seaweed and tolerate the volume expansion and increase the contact of electrolyte [38]. Very recent past, Liang et al. proposed P@PMCNFs/CNTs in which a, flexible P/C electrode by encapsulating P in a dual-conducting network of porous multichannel carbon nanofibers and in-situ carbon nanotubes. This freestanding electrode demonstrates admirable rate capability of 601 mAh g^{-1} at 3 A g^{-1} [39]. Addition conductive material such as carbon nanotube into CNFs can offer a fast-electronic conductivity, thereby enhanced electrochemical behaviour. Recent approach says that the design of brush-like Ni/carbon nanofibers/carbon nanotubes (Ni-CNFs-CNTs) multi-layer conductive network have been reported as electrode for LIBs. Figure 2.6e illustrates the SEM images of CNTs grown on CNFs surface with different time of sputtering Ni under the same CVD treatment (30 s) in which the CNT diameter is about 30 nm and Ni particles exist on the end of CNT. The rate capability test of Ni-CNFs-CNTs anode displayed in Fig. 2.6f which indicating a very stable cycle performance [40].

While metal oxide-based anode materials have been extensively investigated for LIBs due to high energy densities and high theoretical capacity. Moreover, it will be a kind of promising alloy-type materials [41]. For example, Tin oxides, such as SnO and SnO_2, are delightful anodes as an alternative for the commercial graphite as their theoretical capacity of tin oxides (SnO_2: 783 mAh/g, SnO: 875 mAh/g) has been assessed to be outstanding to that of graphite. Nonetheless, its practical use is troubled by a poor cycle that is triggered by the mechanical impairment of the electrodes due to a large expansion (ca. 300%) with alloying while the Li^+ insertion process [42]. Some evolution has been prepared to fabricate free-standing electrodes using SnO_x particles that are distributed onto the carbon support and achieved remarkable capacity [43]. Xia et al. reported SnO_2@N-CNF free-standing film in which ultra-fine SnO_2 nanoparticles finely dispersed in nitrogen-doped carbon nanofibers. As prepared SnO_2@N-CNF electrode still unveiled a reversible capacity of 754 mAh g^{-1} with nearly 100% coulombic efficiency. The corresponding reversible reaction between the voltage of 0.01–3 V can be expressed as follows [44]

$$SnO_2 + 4Li^+ + 4e^- \rightarrow Sn + 2Li_2O \tag{2.7}$$

$$xLi^+ + Sn + xe^- \leftrightarrow Li_xSn(0 < x \leq 4.4) \tag{2.8}$$

$$C + Li^+ \leftrightarrow Li_xC \tag{2.9}$$

Diversely, Zhang et al. C/Fe_2O_3/CNF in which graphitic carbon-coated Fe_2O_3 nanoparticles embedded in carbon nanofibers by one-pot electrospinning method. While a graphitic carbon layer was in situ deposited on the Fe particle surface. The TEM images of C/Fe_2O_3/CNF composites shown in Fig. 2.7a. After annealing, the Fe_2O_3 particles tend to aggregate due to the partial loss of carbon between them, and potential curves in Fig. 2.7b represented a clear discharge plateau at about 0.75 V in the C/Fe_2O_3/CNF electrode, comparable to that detected in the well-ordered

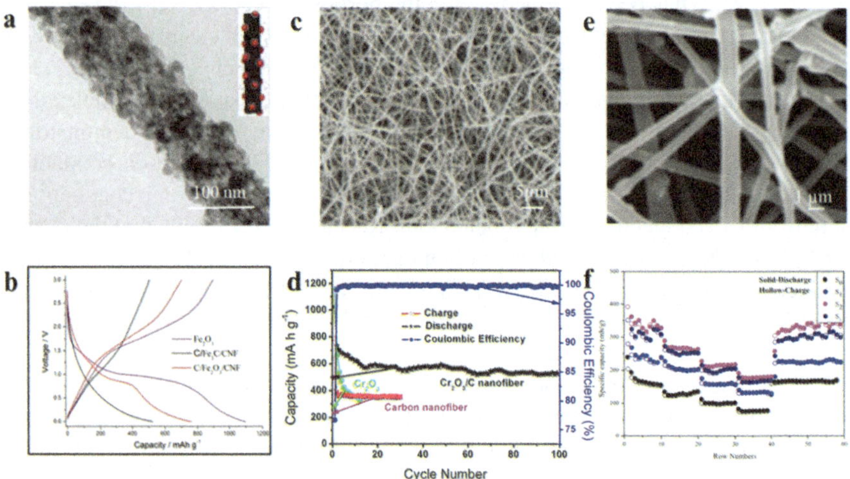

Fig. 2.7 **a** TEM images of C/Fe_2O_3/CNF composites. **b** Charge/discharge curves of the 2nd cycle at 0.1 A g^{-1}. Reproduced from Ref. [45]. Copyright 2014, The Royal Society of Chemistry. **c** SEM images of Cr_2O_3/C nanofibers. **d** Cyclic performance of Cr_2O_3/C nanofiber, Cr_2O_3 nanoparticles and carbon nanofiber at 0.1 A g^{-1}. Reproduced from Ref. [47]. Copyright 2016, Elsevier. **e** SEM images of prepared TiO_2@CNF (S_2). **f** Rate capabilities of S_0, S_1, S_2 and S_3 electrodes at various currents. Reproduced from Ref. [49]. Copyright 2018, Elsevier

Fe$_2$O$_3$ electrode, but with a lower capacity [45]. Likewise, Joshi et al. reported FeO$_x$-CNT/CNF nanocomposite mats to buffer the volume expansion of FeO$_x$ and assure high coulombic efficiency. Furthermore, the integration of FeO$_x$ and CNTs with 2 wt% FeO$_x$-CNT/CNF resulted in superior rate performance of 580 at 1000 mA g^{-1}, because of the uniform spread of the CNTs and FeO$_x$ particles on the CNFs [46]. Besides, Yang et al. has chosen Cr$_2$O$_3$ as the active material to modify the carbon nanofibers as it holds a high theoretical capacity and unveils comparative low electromotive force value about 1.08 V versus Li$^+$/Li. Figure 2.7c manifests the low magnification SEM image of as-prepared Cr$_2$O$_3$/C nanofibers, in which super long fibers with uniform diameter approximate 150 nm were observed. Furthermore, this freestanding electrode disclosed a high specific capacity and stable cyclic performance of 527 mAh g^{-1} in the 100th cycle displayed in Fig. 2.7d [47]. Subsequently, MnO based free-standing electrode has been developed by using electrospinning technique and achieved good structural stability and high surface area. For instance, Wang et al. synthesized the freestanding and flexible core–shell MnO/carbon nanofiber (CNF) composite thin films by host–guest interaction and high-voltage electrospinning technique. This structure has greater accessibility to active sites for the Li$^+$, shorter diffusion paths and rapid Li$^+$ movement. Thus elucidating that the unique core–shell carbon layer MCNFs material structure can donate the electrode with the stable repetitive capability and greatly enhance Li$^+$ conveyance during the electrochemical reaction [48]. Another noteworthy metal oxide is titanium dioxide (TiO$_2$) that receives an industry deliberation as metal oxides due to its inherent properties like low cost, good structural stability and safe working potential. TiO$_2$ with anatase is advantageous as a Li$^+$ holder among other several crystal structures of TiO$_2$ on account of its open structure for Li$^+$. Yet, the submission of TiO$_2$ is obstructed by its dreadful Li$^+$ diffusivity and electrical conductivity analogous to other transition metal, which origins inadequate specific capacity and rate performance as anodes for LIBs. Hence Liang et al. prepared a free-standing film of TiO$_2$@CNFs by electrospinning method. The nitrogen content and the pore in the composite are controlled by diisopropyl azodiformate (DIPA) addition instead of old-style urea and polyvinylpyrrolidone as a sacrifice. The SEM images of sample S$_2$ shown in Fig. 2.7e corresponding to the weight of DIPA was 5 g. The rate performance for the samples was examined. Deceptively, the addition of DIPA meaningfully improved the electrochemical performance from the evaluation of the other samples provided in Fig. 2.7f. Afterwards, there are several metal oxides based free-standing electrode have been discovered to attain some high electrochemical characteristics [49]. Among them, V$_2$O$_3$/CNF membrane offers a highly conductive pathway for rapid charge-transfer reactions, impressively enhancing the active surface sites of V$_2$O$_3$ for fast lithiation/delithiation of Li$^+$ [50]. Excitingly, binary metal oxides such as SnO$_x$–ZnO [51], Fe$_2$O$_3$–SnO$_x$ [52] and TiO$_2$–SnZnO [53]. Two metal oxide phases react with Li at different potentials; therefore, volume expansion and contraction occur in two stages, eventually dropping the strain and improving the stability. During the last few years, several composites have been devoted to finding a suitable freestanding electrode and successfully demonstrated their electrochemical features. Indeed many hybrid composites namely MnO$_2$–Mo$_2$C [54], Ni@ZnO [55], Fe$_{1-x}$SFe$_3$O$_4$ [56] and

$Co/Co_3SnC_{0.7}$ [57] exhibited superior performance when its incorporated with CNFs. Specifically, they improved structural stability and buffered the huge volume changes when it's undergone long-term cycling.

2.4 Binder-Free Electrodes

2.4.1 Binder-Free Cathodes

$LiCoO_2$ is the primary cathode material in real-world rechargeable LIBs. It has advantages of high structural reversibility, below 4.2V Li/Li^+ and decent capacity retention. The capacity fading of $LiCoO_2$ cathodes is always observed upon prolonged cycling and further increases in the charge cut-off voltage, especially at elevated temperatures [58]. Considerable effort has been made to explain the precise mechanism underlying the capacity fading observed at high charge potentials. Research on binder-free cathode for use in LIBs are showing remarkable interest since the insulating binders reduce the overall energy density by adding weight to the electrodes and lead to poor electron transfer on cycling. A few years ago, Luo et al. developed $LiCoO_2$-SACNT binder-free cathode composite an ultrasonication and co-deposition (USCD) technique resulting high conductivity, great flexibility, and outstanding cycling stability (151.4 mAh g^{-1} at 0.1 °C with retention of 98.4% over 50 cycles. The continuous and three-dimensional SACNT network renders well-developed porous structures, high mechanical strength, and superior flexibility of the composites [59]. Consequently, Toprakci et al. prepared $LiFePO_4$/CNT/C composite. Herein Functionalized CNTs were used to increase the conductivity of the composite. Schematic diagram of $LiFePO_4$/CNT/C composite nanofibers shown in Fig. 2.8a. $LiFePO_4$/CNT/C composite nanofibers kept their "networklike" structures after heat treatment with the fiber diameter of 16 nm shown in SEM image see Fig. 2.7b. The cycling performance says that the reversible capacities remain relatively constant for $LiFePO_4$/C and $LiFePO_4$/CNT/C composite nanofibers (i.e., 161 and 169 mAh g^{-1}, respectively) over the full fifty cycles. These standards correspond to 95 and 99% of the theoretical capacity of $LiFePO_4$ [60] illustrated in Fig. 2.7c.

2.4.2 Binder-Free Anodes

The consumption of polymeric binders during electrode construction will inexorably decline the electrical conductivity, reduce the accessibility of active materials and experience kinetic complications. As a consequence, the electrode cannot preserve the reliability and suffer from pulverization, which effects in poor cycling stability

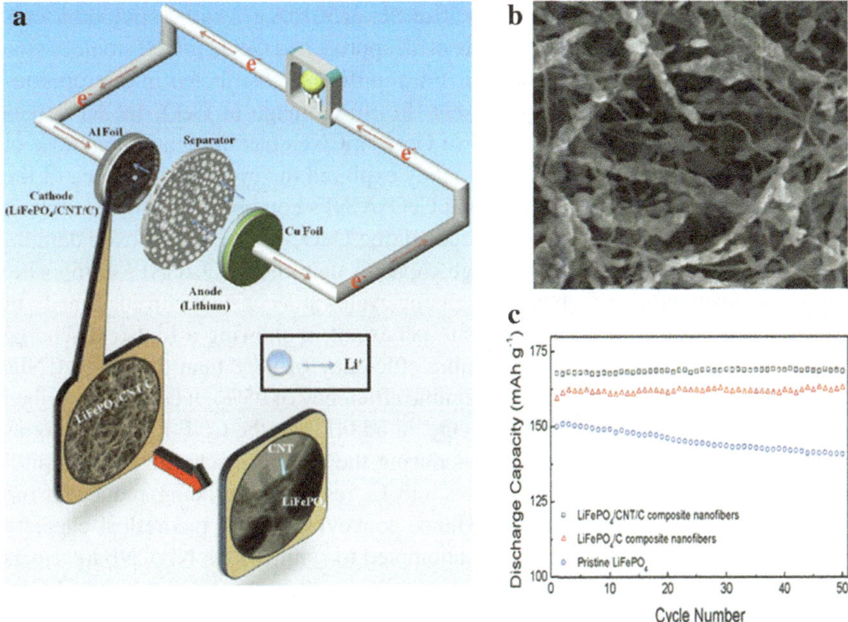

Fig. 2.8 **a** Schematic diagram of LiFePO$_4$/CNT/C composite nanofibers. **b** SEM images of pristine LiFePO$_4$ powder, LiFePO$_4$/C composite nanofibers. **c** Cycling performance of pristine LiFePO$_4$ powder, LiFePO$_4$/C LiFePO$_4$/CNT/C composite nanofibers. Reproduced from Ref. [60]. Copyright 2012, American Chemical Society

and sluggish rate capability [61]. In the past years, several studies have been directed on binder-free electrode materials as an anode for rechargeable Li-ion batteries [62–65]. In general, Germanium is a favourable high-capacity anode material for LIBs but still suffers from poor cyclability owing to its huge volume expansion during the Li–Ge alloy/de-alloy process. Therefore, Li et al. reported Ge-CNF flexible film as binder-free electrodes for LIBs. Remarkably, the Ge-CNFs film reveals superb electrochemical characteristics with a reversible specific capacity of 1420 mAh g^{-1} over 100 cycles at 0.15 °C with only 0.1% decay per cycle (the theoretical specific capacity of Ge is 1624 mAh g^{-1} corresponding to Li$_{4.4}$Ge) [66]. Afterwards, Panto et al. synthesized C/GeO$_2$ paper-like membrane and utilized as a binder and collected free anodes in LIBs. In these studies, they stated that the size of the GeO$_2$ NPs and the fiber diameter increase as increasing germanium load. Finally, the observed capacities of the composite fibers are expected to range between 590 and 780 mAh g^{-1} [67]. Subsequently, the same research group have been developed a new strategy to improve the C/GeO$_2$ paper-like electrode by incorporating Ge nanoparticles via electrospinning and carbonization process. As advanced C/Ge/GeO$_2$ composites also act as binder and collector free anode for LIBs. Specifically, the membrane prepared from solution with 4.25 wt% Ge-load by cold-pressing and carbonization at 700 °C, can deliver about 1500 mAh cm^{-3} after 50 cycles at 50 mA g^{-1} with a coulombic efficiency close to 100% [68]. Likewise, He et al. approached GeO$_x$-mCNFs binder-free electrode in which amorphous GeO$_x$ is ultra-uniformly distributed in microporous carbon nanofibers. Figure 2.9a represent the SEM image of GeO$_x$-mCNFs. It is deceptive that the surface morphology of GeO$_x$-mCNFs has smoother than those of the GeO$_x$/CNFs. As shown in Fig. 2.9b they explored the cycle performance of the CNFs matrix, and the GeO$_x$-mCNFs and GeOx/CNFs composite electrodes under a current density of 1.2 A g^{-1}. As anticipated, the GeO$_x$-mCNFs composite demonstrated a higher second-cycle discharge capacity than the GeO$_x$/CNFs composite. Undeniably, even after 300 charge–discharge cycles at a constant current density of 1.2 A g^{-1}, the GeO$_x$-mCNFs composite persistent in showing a higher discharge capacity of 621 mAh g^{-1} and coulombic efficiency of 98% than the GeO$_x$/CNFs composite (i.e., 453 mAh g^{-1} and coulombic efficiency of 95%). It could be ascribed that the ultra-uniform dispersion of GeO$_x$, in addition to the CNF matrix acting as a buffer against GeO$_x$ volume changes during the charge–discharge process [69]. Metallic Sn is not compulsory to consume Li resource for being reduced from oxides (e.g. SnO$_2$) in the lithiation. Hence conveying higher theoretical capacity than its oxides (SnO$_2$). So Wang et al. attempted to synthesis Sn NP/CNF hybrid as

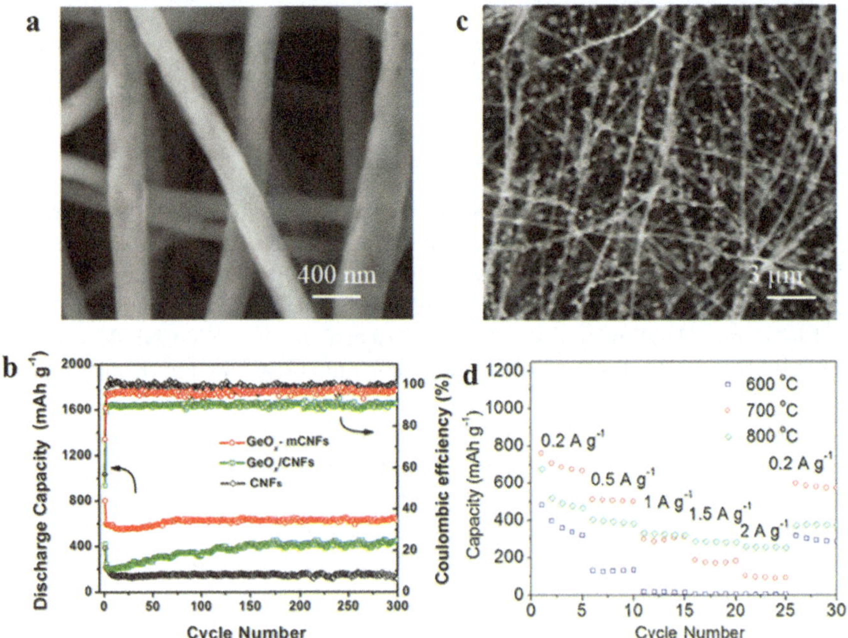

Fig. 2.9 **a** SEM image of GeO$_x$-mCNFs. **b** Discharge capacities and corresponding coulombic efficiencies with respect to cycle number of GeO$_x$-mCNFs measured at a current density of 1.2 A g^{-1}. Reproduced from Ref. [69]. Copyright 2017, Elsevier. **c** Sn NP/CNF hybrids carbonized at 800 °C. **d** Rate capabilities of the samples carbonized at 600, 700 and 800 °C. Reproduced from Ref. [70]. Copyright 2014, Elsevier

the binder-free electrode for use in LIBs. This SEM image of the hybrid is presented in Fig. 2.9c. The SEM images recommended that increased amount of Sn N.P.s were precipitated on the surface of CNFs. The corresponding rate performance of all samples prepared at different calcination temperature illustrated in Fig. 2.9d [70].

Although two dimensional (2D) layered nanostructures have prompted increasing research actions in recent years due to their unique physicochemical properties for a large scale of possible applications from energy storage and catalysts to electronics. Beyond the 2D layered structures are more affirmative for alleviating volume change compared with the bulk counterparts [71, 72]. In contrast, 2D chalcogenide materials consisting of non-oxide elements (S, Se and Te) have recently been examined as new electrochemical active components. Among them, SnSe has been recognized as a promising anode material with a high theoretical capacity of 847 mAh g^{-1}. Sn-based materials such as metallic Sn and SnO_x, are alternative substances for anode materials due to their high theoretical specific capacities (990 mAh g^{-1} for Sn, 1273 mAh g^{-1} for SnO, and 1494 mAh g^{-1} for SnO_2). Yet, the large volume changes and subsequent pulverization during the repeated lithiation/delithiation processes cause poor cycling stability. To rectify these issues, many strategies have been addressed. Markedly, Yuan et al. designed SnSe/SnO_x@CNFs) as a binder-free anode in which of SnSe nanorods embedded in SnO_x@CNFs. The SEM images of 20-SnSe/SnO_x@CNFs, tiny and negligible nanoparticles would separate from the nanofibers, which can be seen in the dotted area shown in Fig. 2.10a. Further, the optimal SnSe/SnO_x@CNFs exhibited a higher initial coulombic efficiency (86.5%) and better reversible capacity (740.7 mAh g^{-1} at 200 mA g^{-1} after 70 cycles) than pristine SnO_x@CNFs (68.1%, 552 mAh g^{-1} at 200 mA g^{-1} after 70 cycles) represented in Fig. 2.10b [73]. Similarly, 2D SnS_2 materials are talented for energy storage, but the specific capacity and cycling performance are still far from adequate. Thus, Wang et al. designed SnS_2 nanosheets array on a low-cost carbon paper as the binder-free electrode through the hydrothermal method. The TEM image of SnS_2@C hybrid paper is shown in Fig. 2.10c and the thickness of the nanosheets is valued to be about 10–30 nm. The galvanostatic charge/discharge curves of the SnS_2@C and SnS_2 anodes at a current density of 100 mA g^{-1} demonstrated in Fig. 2.10d. The charge/discharge plateaus are in well-matched with the Li-ion reactions in the CV experimental results. It can be seen that the SnS_2@C anode delivers initial discharge and charge capacities of 1851 and 1156 mAh g^{-1}, respectively, which are much higher than that of the pure SnS_2 anode (i.e., 1304 and 755 mAh g^{-1}) [74]. Also, a recent report says that the intermetallic collected of two or more metals may expose less volume discrepancy due to the different redox potentials of the metals during lithiation/delithiation processes, which can dramatically improve the electrochemical performances. For example, Chen et al. SnSb-CNTs@NCNFs) as a binder-free and current-collector-free anode with outstanding electrochemical performances and flexibility for LIBs. The SnSb nanodots are well distributed in the NCNFs and thus evade agglomeration during continuous cycling. The addition of CNTs can commendably enhance the conductivity and robustness of the electrode. The general Li-storage reactions of SnSb-CNTs@NCNFs can be summarized as follows [75]

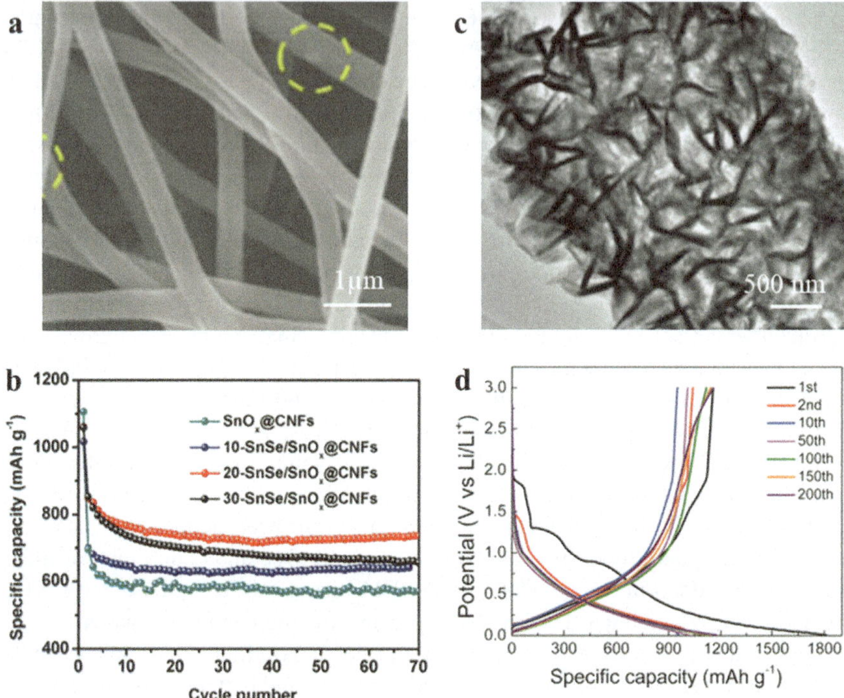

Fig. 2.10 **a** SEM images of 20-SnSe/SnOx@CNFs. **b** Cycling performances of all samples at 200 mA g^{-1}. Reproduced from Ref. [73]. Copyright 2018, The partner Organizations. **c** TEM images of SnS$_2$@C composites. **d** Galvanostatic charge/discharge curves. Reproduced from Ref. [74]. Copyright 2017, The Royal Society of Chemistry

$$SnSb + 3Li \leftrightarrow Li_3Sb + Sn \tag{2.10}$$

$$Li_3Sb + Sn + xLi \leftrightarrow Li_3Sb + Li_xSn(0 \leq x \leq 4.4) \tag{2.11}$$

A novel graphitic carbon-based material such as nitrogen-rich graphitic nanotubes [76] and walnut shell-derived carbon nanofibers [77] have been reported as binder-free electrode for use in LIBs. Noticeable, they have demonstrated great mechanical stability and improved Li-storage, due to the assembly of more Li-storage sites with higher electronegativity.

2.5 Other Composites for Cathodes

It is widely believed that the nanostructured electrodes could expand the energy density and rate capacity and insertion kinetics by diminishing the solid-state diffusion pathway. In past research, there are many studies on the electrospinning fibers for cathode materials, which exhibit superior electrochemical properties compared with the powder materials. Particularly, numerous other electrodes excluding binder and freestanding electrode materials have been addressed, which has demonstrated better electrochemical performance. It includes synthesis of various nanostructured metal oxides [78, 79], carbon coating [80] metal oxide coating [81] and doping [82]. For example, Gu et al. prepared nanostructured $LiCoO_2$ fibers were prepared by the sol–gel associated electrospinning technique using metal acetate and citric acid as starting materials. As-prepared $LiCoO_2$ fibers with 500 nm to 2 μm in diameter were self-possessed of polycrystalline nanoparticles in sizes of 20–35 nm. They revealed a high initial charge and discharge capacity of 216 and 182 mAh/g, respectively [83]. There has been significant interest in emerging electrospun $LiFePO_4$ as a cathode material for high-power, large-scale applications such as electric vehicles [84]. But it has several intrinsic drawbacks including its low ionic and electronic conductivities. To overcome this issues many other electrodes such as $Li[Fe_{1-x}Mn_x]PO_4$ [85], $LiFePO_4/FeS/C$ [86, 87] $LiFePO_4/C/graphene$ [88] and $LiFePO_4/C$ [80] proposed and electrochemical performance were discussed in details. Among them, the electrospun $LiFePO_4/C$ composite has attracted morphology and discharge capacity. Recently, Lee et al. prepared $LiFePO_4/C$ nanofibers through noble Li^+ conducting pathways built along with reduced carbon webs by phosphorus. The phosphidated $LiFePO_4/C$ nanofibers TEM images have shown in Fig. 2.11a with amorphous carbon layers with 10–20 nm in thickness. The initial galvanostatic voltage profiles of the pristine and phosphidated $LiFePO_4/C$ nanofibers illustrated in Fig. 2.11b. The phosphidated $LiFePO_4/C$ nanofibers displayed a high discharge capacity of $\sim$163 mAh g^{-1}, with low-voltage polarization during the first cycle. Although the $LiFePO_4/C$ nanofibers have a well-fabricated 1D nanostructure, the discharge capacity of the pristine $LiFePO_4/C$ is $\sim$127 mAh g^{-1}, which is much lower than the theoretical discharge capacity ($\sim$170 mAh g^{-1}) of $LiFePO_4$ [80]. Meanwhile, the development of $\sim$4 V low cost and eco-friendly spinel phase $LiMn_2O_4$ cathode have been addressed. Excitingly, $LiMn_2O_4$ cathode nanofibers greatly enhanced the long term life cycle [89]. A few years ago, Jayaraman et al. reported porous $LiMn_2O_4$ hollow nanofibers through electrospinning method and stated that the electrospun $LiMn_2O_4$ hollow nanofibers retained 87% of initial reversible capacity after 1250 cycles at the 1 °C rate. Due to the 1D hollow morphology, they could achieve an excellent performance [90]. Later another spinal phase $LiMn_{1.5}Ni_{0.5}O_4$ cathode with 1D hollow and porous nanofibers developed since it can deliver an energy density as high as 658 Wh kg^{-1}, which is much higher than those of $LiMn_2O_4$ (440 Wh kg^{-1}) and $LiFePO_4$ (500 Wh kg^{-1}) [91]. Vanadium oxides are an identical encouraging electrode material for LIBs owing to their layered crystal structure that offers more energy storage

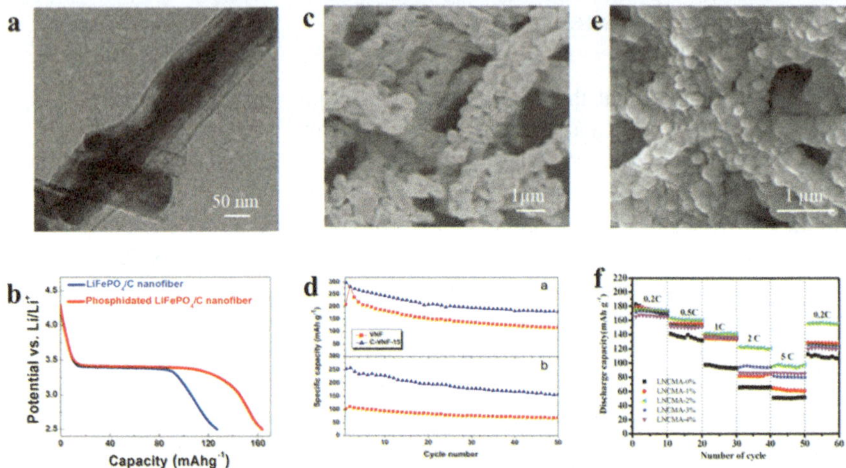

Fig. 2.11 **a** TEM images of phosphidated LiFePO$_4$/C nanofibers. **b** Potential profiles of initial galvanostatic voltage profiles. Reproduced from Ref. [80]. Copyright 2014, American Chemical Society. **c** FESEM images of C-VNF-15. **d** Plots of specific discharge capacity vs. cycle number of VNF and C-VNF-15 cycled at **d(a)** 0.1 °C rate at 551 °C and **d(b)** 1 °C rate at room temperature. Reproduced from Ref. [94]. Copyright 2018, The Royal Society of Chemistry. **e** SEM images of 2% Au-doped Li1.2Ni$_{0.7}$Co$_{0.1}$Mn$_{0.2}$O$_2$ nanofibers. **f** Rate performance of Au-doped Li$_{1.2}$Ni$_{0.7}$Co$_{0.1}$Mn$_{0.2}$O$_2$ nanofibers. Reproduced from Ref. [82]. Copyright 2012, Elsevier

via metal-ion intercalation, in addition to good catalytic activity for energy conversion. A series of investigations on electrospun V$_2$O$_5$ was reported previously, and those report illustrated improved structural stability against volume expansion, thus improving cycling stability [92, 93]. Cheah et al. prepared V$_2$O$_5$ (VNF) nanofibers and decorated with carbon coating by PECED method. The carbon coating was 15 and 30 min as denoted as C-VNF-15 and C-VNF-30, respectively. The FE-SEM images of C-VNF-15 provided in Fig. 2.11c. It is evident that the carbon layer with thickness of 5 nm over V$_2$O$_5$ nanocrystals. In the elevated temperature conditions improved discharge capacity is noted for C-VNF-15 in the first cycle when compared to room temperature conditions (280–298 mAh g^{-1}) see in Fig. 2.11d(a). The high rate cycling performance of VNF and C-VNF-15 between 1.75 and 4.0 V versus Li at 1 °C (350 mA g^{-1}) at room temperature represents in Fig. 2.11d(b).

The decrease of capacity in elevated temperature state is primarily ascribed to the severe reactivity of VNFs with electrolyte counterpart [94]. Very recently, electrospun layered cathode material was reported, and their impact of metal doping effect have been discussed. Yue et al. reported Au-doped Li$_{1.2}$Ni$_{0.7}$Co$_{0.1}$Mn$_{0.2}$O$_2$ nanofibers. SEM photos revealed the electrospun nanofibers have a uniform particle size in the range of 300–400 nm in Fig. 2.11e. The ideal doping amount of Au is 2% in Li$_{1.2}$Ni$_{0.7}$Co$_{0.1}$Mn$_{0.2}$O$_2$ to obtain high discharge capacity and excellent capacity retention as given in Fig. 2.11f. These findings disclose that stable structure with good particle contact of the Au-doped cathode materials can enhance electrochemical properties [82]. Vanadium pentoxide (V$_2$O$_5$) is one of the most encouraging cathode materials since it can capture large quantities of electrons owing to the layered structure, which can deliver an admirable performance during the lithiation/delithiation process. V_2O_5 can bring a high theoretical capacity (294 mAh g^{-1}, vs. Li/Li$^+$), which is a higher value compared to other commercialized electrode materials for a cathode [95]. But, V_2O_5 cathodes undergo a structural change or irreversible transformation beyond 2 mol of Li-insertion, which causes severe capacity fading upon cycling [96]. Therefore, many methods like carbon coating and metallic doping, are endeavoured with a variety of morphologies to expand battery performance [94, 97]. For example, An et al. reported C/HPV$_2$O$_5$ in which carbon encapsulated hollow porous vanadium-oxide nanofibers. Figure 2.12a illustrates the schematic diagram of the ideal C/HPV$_2$O$_5$-30 synthesis process via electrospinning and post calcination treatment. As prepared samples were sintered in air for 10, 30, and 60 min using a box furnace at 400 °C to form carbon on V_2O_5. The C/HPV$_2$O$_5$-30 (Fig. 2.12b) with diameters ranging from 440 to 453 nm showed the hollow porous structure due to the diffusion of V.N., a phase transition from V.N. to V_2O_5, a decomposition of carbon, and grain growth of V_2O_5. The cycling durability was evaluated by galvanostatic charge–discharge tests up to 100 cycles. Noticeably, the C/HPV$_2$O$_5$-30 electrode presented impressive cycling durability (see Fig. 2.12c with a high specific discharge capacity of 241 mAh g^{-1} after 100 cycles with a capacity retention of 82%, which is the most top performance in comparison with previously reported V_2O_5 cathode materials with various morphologies and structure. The exceptional cycling performance attributes to the carbon coating, and porous structure buffers the volume expansion and shorter Li-ion diffusion [95].

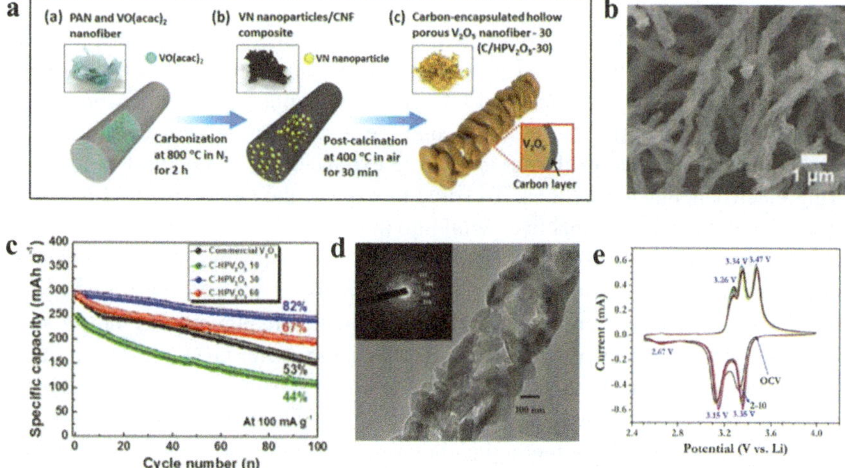

Fig. 2.12 **a** Schematic illustration of the ideal C/HPV$_2$O$_5$-30 synthesis process. **b** Low-resolution SEM images C/HPV$_2$O$_5$-30. **c** Cycling durability of the carbon coated samples. Reproduced from Ref. [95]. Copyright 2016, American Chemical Society. **d** TEM images of sintered VNF at 400 °C. (inset) SAED pattern. **e** Cyclic voltammograms of VNF in half-cell assembly (Li/VNF) between 2.5 and 4.0 V versus Li. Reproduced from Ref. [98]. Copyright 2013, American Chemical Society

Another study proposed morphological features of the electrospun VNF via analyzing through TEM. Deliberate degradation of as-spun fibers during the sintering process results in the retention of fibrous morphology, which is composed of V_2O_5 nanoparticles grown along the fiber direction with porous structure observed from the TEM image (Fig. 2.12d). Following CV test for the VNF in half cell assembly very high open-circuit voltage (OCV) $\sim$3.4 V versus Li and subsequently discharged to 2.5 V versus Li for Li-insertion. From CV traces, it is apparent that VNF half-cells exhibited several oxidation/reduction peaks during both cathodic and anodic scan, which corresponds to reversible insertion and extraction processes of Li in the V_2O_5 lattice. Also, the current density of CV traces stays constant without significant reduction in area under the curve, which indicates nice cycling properties of such VNF in half-cell configuration [98].

Table 2.1 illustrates the various types of electrospun based lithium metal oxides cathode for LIBs. Generally known, the most typical lithium-insertion cathode materials are lithium transition metal oxides, which mainly include $LiMO_2$ (M often refers to Co, Mn, Ni) with a layered structure and LiM_2O_4 with spinel structure. Besides, recent developments in electrospinning technique has seen a frequently expansion of electrospun hollow nanofibers as cathodes for LIBs.

2.6 Other Composites for Anodes

2.6.1 Metal/CNF Composites

Amid several types of efficient nanomaterials, active metals can store a large amount of Li by forming alloys through a Li insertion reaction and hence they are appropriate for use as high-capacity anode materials in rechargeable LIBs [108]. Conversely, these active metals suffer from severe particle accumulation and capacity worsening while cycling due to the large volume expansion and structural destruction during electrochemical reactions. To solve this issue, the simple method has been introduced that dispersing inactive metal into the carbon matrix to form metal-filled carbon nanofibers (M/CNFs) by electrospinning and following thermal annealing procedures. As prepared composite deliver more than 400 mA g^{-1} reversible capacities at 50 and 100 mA g^{-1} current densities and also uphold pure fibrous morphology and good structural integrity after 50 charge/discharge cycles [109]. Furthur reduces the limitation of the metal anode, intermetallic binary systems have been addressed [110, 111]. For example, Jung et al. prepared Ni–Sn/CNF nanocomposites as anode materials by electrospinning method and thermal annealing process. The nanocomposite prepared at 700 °C had disordered Ni_3Sn_2, disordered NiO, amorphous SnO_x and crystalline SnO_2 phases created as tiny particles in the carbon nanofibers. Figure 2.13a indicates the FE-SEM images of Ni–Sn/CNF at 700 °C, which reveals a complete, interconnected structure with an average diameter of about 180 nm.

Table 2.1 Different types of electrospun based lithium metal oxides cathode for LIBs

Electrodes	Precursors	Process	Morphology	LIB performance	References
$0.4Li_2MnO_3 \cdot 0.6LiNi_{1/3}Co_{1/3}Mn_{1/3}O_2$	$PAN/Li(Ac)_2 + Ni(Ac)_2 + Co(Ac)_2 + Mn(Ac)_2/DMF$	Electrospinning	Nanotubes	199 mAh g^{-1} after 50 cycles at 0.1 °C	[99]
$Li_{1.2}Mn0.54Ni_{0.13}Co_{0.13}O_2$	$PAN/Li(Ac)_2 + Ni(Ac)_2 + Co(Ac)_2 + Mn(Ac)_2/DMF$	Electrospinning and two-steps thermal treatment	Mesoporous nanotubes	~140 mAh g^{-1} after 300 cycles at 3 °C	[100]
$LiNi_{0.8}Co_{0.1}Mn_{0.1}O_2$-MgO	$PVP/Li(Ac)_2 + Ni(Ac)_2 + Co(Ac)_2 + Mn(Ac)_2/citric$ acid/H2O	Coaxial electrospinning	Hollow nanofibers	174 mAh g^{-1} after 50 cycles at 0.02 A g^{-1}	[101]
$0.5Li_2MnO_3$-$0.5LiNi1/3Co1/3Mn_{1/3}O$	$PVA/Li(Ac)_2 + Ni(Ac)_2 + Co(Ac)_2 + Mn(Ac)_2/H2O$	Electrospinning	Hollow nanowires	~200 mAh g^{-1} after 20 cycles at 0.01 A g^{-1}	[102]
$LiMn_2O_4$	$PVP/Li(NO_3)_2 + Mn(Ac)_2/$ ethanol	Electrospinning	Hollow nanofibers	103 mAh g^{-1} after 1250 cycles at 0.15 A g^{-1}	[90]
$LiNi0.5Mn_{1.5}O_4$	$PVA/Li(Ac)_2 + Ni(NO_3)_2 + Mn(Ac)_2/ethanol$	Electrospinning	Porous nanofibers	109 mAh g^{-1} after 50 cycles at 0.15 A g^{-1}	[103]
C-coated $LiFePO_4$	$MMA/LiH_2PO_4 + Fe(NO_3)_3 + Fe(SO_3)_2/DMF$-PAN/DMF	Coaxial electrospinning and two-steps thermal treatment	Hollow nanofibers	135 mAh g^{-1} after 20 cycles at 2 °C	[104]
$LiFe_{1-y}Mn_yPO_4/C$	$PVP/LiOH + Mn(SO_3)_2 + Fe(SO_3)_2/H2O$	Electrospinning	Hollow nanofibers	~100 mAh g^{-1} after 50 cycles at 4 °C	[105]

(continued)

Table 2.1 (continued)

Electrodes	Precursors	Process	Morphology	LIB performance	References
LiFePO$_4$/C/Ag	PVP/LiOH + Fe(NO$_3$)$_3$ + H$_3$PO$_4$ + AgNO$_3$/DMF	Coaxial electrospinning and two-steps thermal treatment	Hollow nanofibers	138 mAh g^{-1} after 0.2 °C	[106]
V$_2$O$_5$	PVP/vanadium acetylacetone/DMF	Electrospinning	Porous nanotubes	105 mAh g^{-1} after 250 cycles at 2 A g^{-1}	[107]
C-encapsulated V$_2$O$_5$	PAN/VO(acac)$_2$/DMF	Electrospinning	Porous nanotubes	241 mAh g^{-1} after 100 cycles at 0.1 A g^{-1}	[95]

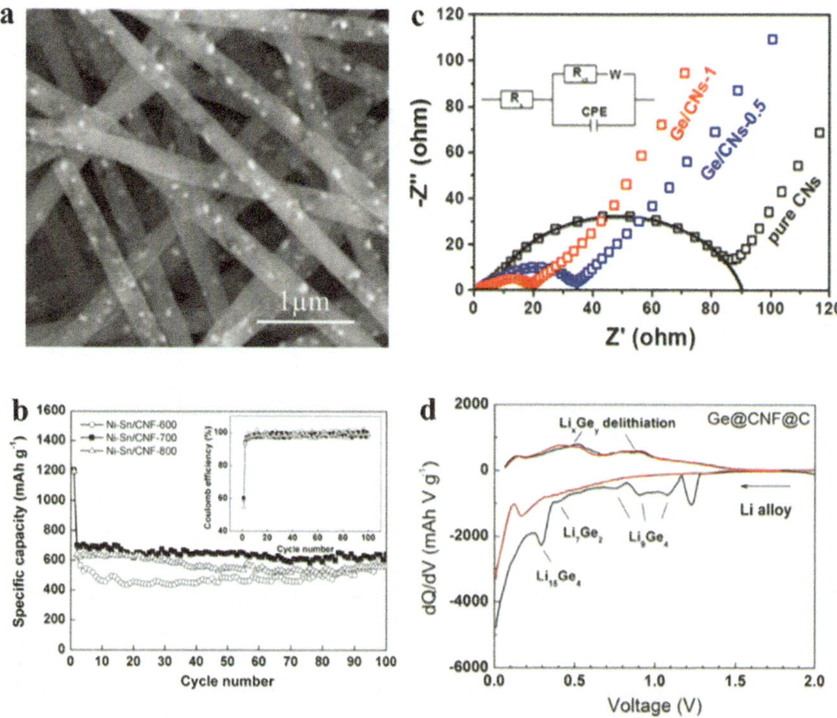

Fig. 2.13 **a** SEM images of Ni–Sn/CNFs at 700 °C. **b** Cycle performance and coulomb efficiency of Ni–Sn/CNFs with various temperatures at 0.5 mA cm^{-2} in 1 M LiPF$_6$/EC/DMC. Reproduced from Ref. [112]. Copyright 2011 The Electrochemical Society. **c** Nyquist plots of impedance data of Ge/CNFs. Reproduced from Ref. [113]. Copyright 2016, American Chemical Society. **d** Differential capacity plots of the first and second cycles of Ge@CNF@C. Reproduced from Ref. [114]. Copyright 2013, Elsevier

Cycle performance and coulomb efficiency of Ni–Sn/CNFs with various temperatures at 0.5 mA cm^{-2} in 1 M LiPF$_6$/EC/DMC displayed in Fig. 2.13b. Ni–Sn/CNF at 700 °C sample delivers a discharge capacity of 641 mAh g^{-1} and its ascribed to the formation of Ni$_3$Sn$_2$ intermetallic compounds and the buffering role of the CNF [112]. Germanium is group IV material that considered as the most promising alternative electrode materials, because of their extremely high theoretical capacity of 1600 mAh g^{-1}. To date, a lot of research has been concentrated on the progress and scheme of novel Ge-based complex nanostructures to expand structural stability during Li alloying/dealloying processes. Ge electrodes exhibits of huge volume change (>300% for fully lithiated states). Ge/carbon nanostructures offer excellent electrochemical abilities due to their high electrical conductivity see in Fig. 2.13c and unique nanostructures are containing well-embedded Ge NPs [113]. Another studies by Li et al. stated that in the Ge@CNF@C composite the Li alloying process, peaks indicate the Li insertion into equipotential sites, and the occurrence of multiple sharp peaks at the first cycle suggests the formation of a number of different Li$_x$Ge$_y$ phases during electrochemical lithiation. In Fig. 2.13d the voltage values (0.50–1.10, 0.30–0.50, 0.15–0.30, and 0–0.15 V) are consistent with those reported for Li_9Ge_4, Li_7Ge_2, $Li_{15}Ge_4$, and $Li_{22}Ge_5$ during Li insertion [114]. Simultaneously, many metal–carbon composites such as Si/C [115], SiO_2/Sb@CNF [116], Sn/C [117] and Sn-SnSb/C [118] have been proposed to overcome the volume expansion issue of the anode materials, and they observed meaning a full reduction in the volume expansion resulting in better electrochemical results compared to the pure metals.

2.6.2 Metal Oxides/CNF Composites

Tremendous research in metal oxide as an anode material for LIBs because of its high Li^+ storage capacity when compared to carbon. Specifically, metal oxides with one-dimensional nanostructure are of special concern as anode materials owing to their high specific surface area, structure-directed electron conduction mechanism [119]. Electrospinning method is simple and versatile for making fibular mesostructures, originates from the fabrication of polymer fibers and has been stretched to the fabrication of many cutting-edge functional nanomaterials [120, 121]. So far, various metal oxides have been prepared by electrospinning process. For example, SnO_2 [122, 123], TiO_2 [124–127], NiO [128, 129], CuO [130] have been studied more intensively and their electrochemical performance for LIBs reported. For example, Aravindan et al. successfully made an attempt to synthesize one dimensional NiO nanofibers by electrospinning technique and examined as anode material in native form for the first time. Figure 2.14a, b displays the photographs of as-spun NiO nanofibers, inset sol–gel precursor used for spinning. The surface morphology studies also carried out to confirm the existence of nanoscopic NiO particles in a continuous fibrous morphology and illustrated in Fig. 2.14c which obviously exemplify the presence of nanosized particles formed as a continuous fibrous morphology and HR-TEM pictures undoubtedly showed the formation of single-crystalline NiO particles. From electrochemical test analysis, found extreme reversible uptake of 1.62 mol of Li after 100 galvanostatic cycles at the current density of 80 mA g^{-1} with a good capacity retention of over 75% of reversible capacity shown in Fig. 2.14d [128].

Also Tin dioxide, as a favourable anode material for LIBs, has drawn immense part on account of its high capacity (781 mAh g^{-1}) compared with graphite (372 mAh g^{-1}). Several methods have been devoted to the synthesis of different nanostructured SnO_2 still many challenges are existed such as volume expansion. Recently, fiber-in-tube and tube-in-tube materials have been discovered since they could maintain their morphology even after repeated lithium insertion and desertion processes [131, 132]. For instance, Hong et al. SnO_2 nanotubes with a fiber-in-tube structure have been prepared for the first time and the preparation scheme illustrated in Fig. 2.15a. The products thus acquired are anticipated to have a comprehensive range of applications in energy storage devices. Figure 2.15b depicts the SnO_2 nanotubes with a fiber-in-tube structure obtained by heating to 500 °C with a diameter of 85 nm were seen to be located inside hollow nanotubes with an outer diameter of 260 nm.

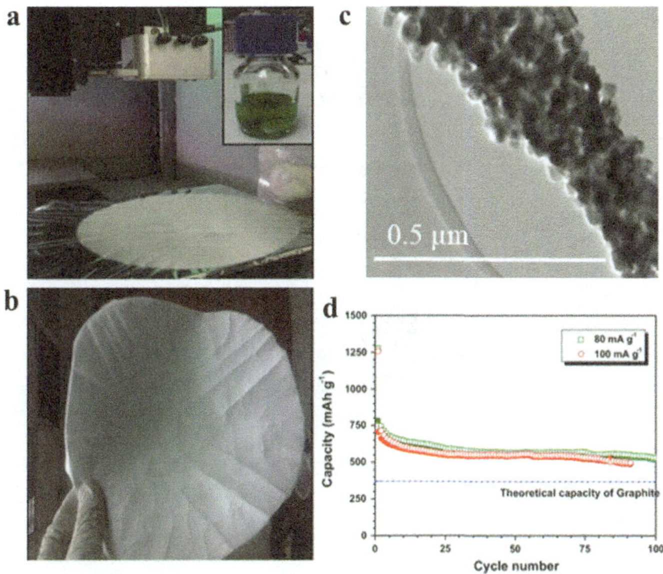

Fig. 2.14 a, b Photographic image of as-spun NiO nanofibers, inset sol–gel precursor used for spinning. **c** TEM pictures of sintered NiO nanofibers. **d** Plot of capacity versus cycle number for Li/electrospun NiO fibers at current densities of 80 and 100 mA g^{-1}. Reproduced from Ref. [128]. Copyright 2012, Elsevier

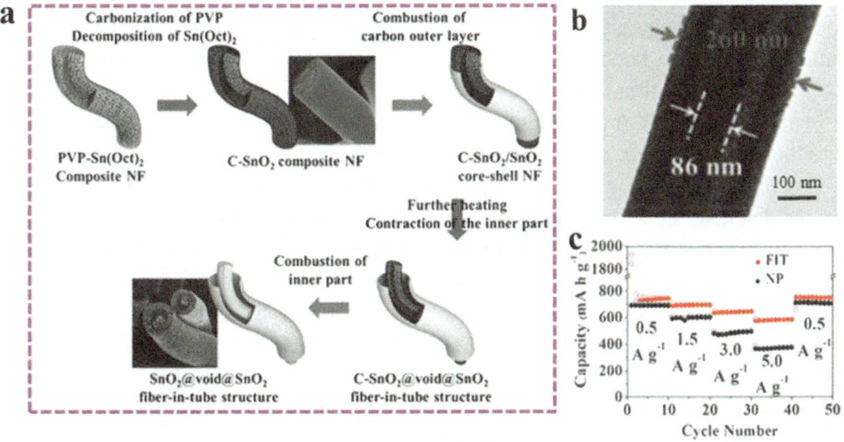

Fig. 2.15 a Schematic depiction of the formation of the SnO$_2$ fiber-in-tube nanostructure. **b** TEM images of SnO$_2$ nanotubes with a fiber-in-tube structure obtained by heating to 500 °C. **c** Rate performances of SnO$_2$ nanotubes and SnO$_2$ nanopowder. Reproduced from Ref. [123]. Copyright 2015 Wiley–VCH

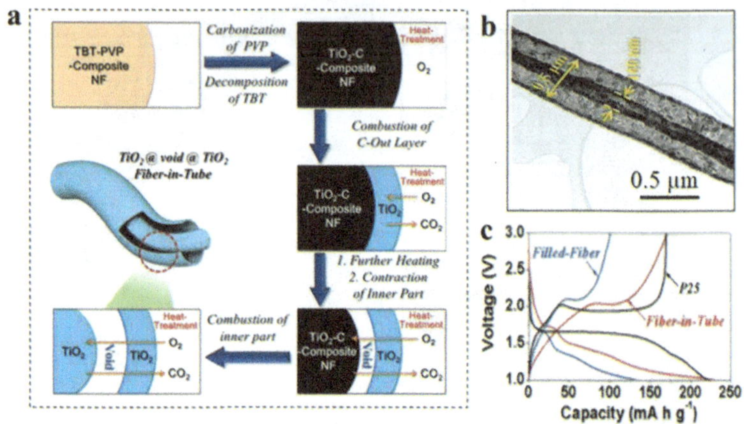

Fig. 2.16 **a** Schematic depiction of the formation mechanism of the TiO_2 nanofiber with fiber-in-tube nanostructure from the TBT-PVP composite nanofiber. **b** TEM image of TiO_2 nanofiber. **c** Initial charge–discharge curves. Reproduced from Ref. [131]. Copyright 2015 Wiley–VCH

Combustion of the SnO_2-carbon composite nanofibers produced SnO_2 nanotubes with a fiber-in-tube structure, which showed high initial discharge capacity as well as excellent cycling properties. The discharge capacities of the SnO_2 nanotubes at incrementally increased current densities of 0.5, 1.5, 3, and 5 Ag^{-1} were 774, 711, 652, and 591 mAh g^{-1}, respectively [123] see in Fig. 2.15d. Moreover, this sort of synthesis strategy opens up for fabrication of various types of anode materials. Almost similar studies had been reported by Cho et al. and proposed phase-pure anatase TiO_2 nanofibers with a fiber-in-tube structure via electrospinning process followed by heat treatment techniques. The formation of mechanism has been explained with the schematic diagram, as shown in Fig. 2.16a.

Calcination is the key factor for the formation of fiber-in-tube structure. The TEM images disclose the formation of a TiO_2 nanofiber with the fiber-in-tube nanostructure. The sizes of the nanofiber core and hollow nanotube measured from the TEM image are 140 and 500 nm, respectively see in Fig. 2.16b. The initial discharge capacities of the TiO_2 nanofibers with fiber-in-tube and filled structures and of the commercial TiO_2 nanopowders were 231, 134, and 223 mAh g^{-1}, respectively, and their core nanofibers with a fiber-in-tube structure improved well to 167 mAh g^{-1} as the current density returned to 0.2 Ag^{-1} after 60 cycles see in Fig. 2.16c. In order to evade the capacity fading of metal oxide anodes, their nanostructuring has been suggested [133]. Carbon decoration, including carbon nanofibers, N-doped carbon nanofibers and graphene, on metal oxides are prooved better performance in terms of post ponding the capacity fading [134–136]. Recently, C-SnO_2 [134], Sn-SnO_x/C [137], $Li_4Ti_5O_{12}$@C@G [138], TiO_2-RGO [139], FeO_x/CNFs [140], Fe_2O_3@C CNFs [141], GN@C/Fe_3O_4 [142], MoO_2@NC [143], MnO@CFs [144], Co_3O_4/CNF [145] and SiO_x/CNF [146]. Another hand hybridizing metal oxide and metal with carbon network have been suggested to control the volume expansion during the electrochemical reactions. Specifically, these carbonaceous materials not only serve as buffer matrix to alleviate the large volume changes of Li-Metal alloying/dealloying but also can bring extra Li storage capacity and increase the electrode conductivity [147, 148]. Likewise, more than one transition metal oxides or mixed oxides have also been considered in recent times. Curiously, a combination of one or more electrospun based metal oxides meaningly delivered good morphological and cycling behaviour. So far, $ZnCo_2O_4$ [149], $CaSnO_3$ [150], $NiTiO_3$ [151] $SrTi_{1-x}Fe_xO_3$ [152] $CaCo_2O_4$ [153], $ZnMn_2O_4$ [154, 155], $CoFe_2O_4$ [156] and $CoMn_2O_4$ [157] nanofibers have fabricated and explored a alternative anode candidate for LIBs. Among them, Co-based oxides have been extensively studied for application in LIBs, because of their high theoretical capacity (ca. 890 mAh g^{-1}), controllable size, and morphology. Also, cobalt is toxic and expensive thus severe efforts have been made to replace Co by eco-friendly and cheaper alternative elements (e.g., Zn, Cu, Ni, Mg, and Fe) without sacrificing the electrochemical performance.

For example, Li et al. produced hierarchical $CaCo_2O_4$ nanofibers via electrospinning method. The effect of the temperature on the morphological and crystallographic phase evolution was discussed. Figure 2.17a represents the preparation scheme of 1D $CaCo_2O_4$ nanostructures calcinated at different temperature. Considerably, the morphologies of the as-prepared $CaCo_2O_4$ 1D nanostructures can be transformed from a nanoplates-in-nanofibers structure to hierarchical nanofibers and nanoplates. The calculated thickness of the nanoplates is around 6 nm see in Fig. 2.17b. Consequently, this structure exhibited admirable cycling performance of 650 mAh g^{-1} after 60 cycles at a current of 100 mA g^{-1} [153] shown in Fig. 2.17c the significant electrochemical performance may be ascribed to the novel 1D hierarchical structure

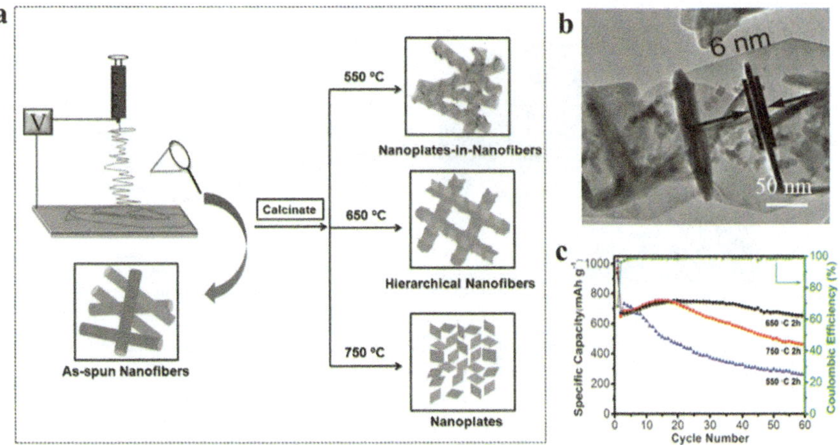

Fig. 2.17 **a** Schematic illustration for the preparation of 1D $CaCo_2O_4$ nanostructures calcination at different temperature. **b** TEM images of $CaCo_2O_4$ calcinated at 650 °C for 2 h. **c** Cycling performance of $CaCo_2O_4$. Reproduced from Ref. [153]. Copyright 2013 Wiley–VCH

which enables fast Li-ion transport. When this kind of oxides mixed or encapsulate with CNFs, effectively prevents the particle pulverization, thereby increased the overall Li-transport during the electrochemical reaction [158]. Hence, Zhou et al. proposed $NiCo_2O_4$ nanoneedles/carbon nanofibers in which tufted $NiCo_2O_4$ nanoneedles directly grown on carbon nanofibers as schematically represented in Fig. 2.18a. The observed length of $NiCo_2O_4$ acicular leaf is around 0.5–2 μm with the width less than 100 nm shown in Fig. 2.18b. The rate performance of both materials is also deliberated at various current densities from 200 mA g^{-1} to 2000 mA g^{-1} as exposed Fig. 2.18c. As the current density increases, a constant decay of specific capacity occurs on both materials. But when recovered back to 200 mA g^{-1}, the TNN/CNFs composite still distributes an average discharge specific capacity above 800 mAh g^{-1} [159].

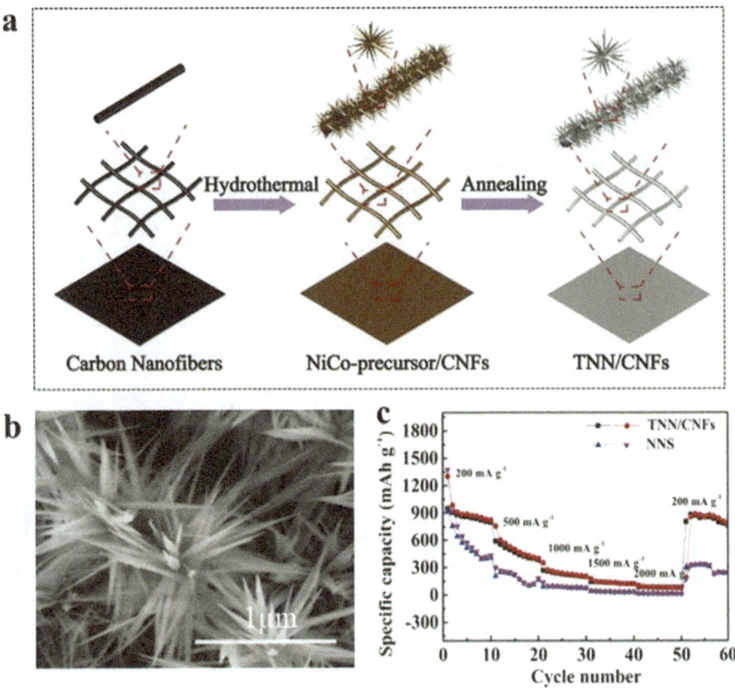

Fig. 2.18 **a** Schematic preparation route for TNN/CNFs. **b** The SEM morphology of tufted $NiCo_2O_4$ nanoneedles/carbon nanofibers (TNN/CNFs) composite TNN/CNFs. **c** Rate performances of TNN/CNFs and NNS composites as a function of current density. Reproduced from Ref. [159]. Copyright 2016, Elsevier

2.6.3 TMD/CNF Composites

Transition metal dichalcogenides (TMDs) have recently appealed vast concern due to their attractive physicochemical properties, principally when they are confined to few layers or to a single layer [160]. A general formula is MX_2 (M = W, Mo, V; X = S, Se, Te), a sandwich structure in which the X–M–X layers are covalently bonded between in-plane atoms and coupled with weak van der Waals forces vertically. This kind of responsiveness can be credited to the exciting phenomenon that a reduction in the dimensionality leads to a variety of striking characteristics which fluctuate from bulk counterparts [161]. MoS_2 with a sandwich structure has gained widespread notice for its hierarchical nanostructure and physicochemical property. Moreover, an excessive number of active sites are dispersed among the layered MoS_2, which provide the ultrathin layered MoS_2 with a high theoretical capacity ($\sim$670 mAh g^{-1}) that is more promising than the commercial electrode materials of LIBs.

Though, its innate poor conductivity (relating to the electron and Li$^+$ transfer) and frail sulfur electrochemistry lead to severe capacity fading and poor rate performance. Numerous approaches have been dedicated to resolving the above problems of MoS_2. It has been verified that a comparatively extensive distance between the atom layers is beneficial to the Li$^+$ transfer to form Li_2S on MoS_2, which accelerates the lithiation/delithiation processes and reduces the destruction of MoS_2 nanostructure during the electrochemical process. For example, Zhu et al. reported single-layered ultrasmall nanoplates of MoS_2 embedded in thin carbon nanowires by electrospinning fo the first time. When they use it for LIBs, the observed capacity is 661 mAh g^{-1} even after 1000 cycles at a high current density of 10 A g^{-1} [162]. Similarly, Zhang et al. fabricated PCNF@MoS_2 by using the strategy of assembling MOFs in PAN electrospun fibers. The hierarchically structured PCNF@MoS_2 fibers via electrospinning and subsequent a hydrothermal process, in which MoS_2 nanoplates were vertically put together on the surfaces of PCNFs represented in Scheme Fig. 2.16a. The corresponding SEM images of PCNF@MoS_2 shown in Fig. 2.16b which reveals that the synthesized MoS_2 nanoplates were almost entirely covered and were uniformly vertically assembled on the PCNFs with no restacking. The diameter of PCNF@MoS_2 was increased from 500 to 600–700 nm at the same time. The potential profiles for PCNF@MoS_2 indicated an ultrahigh initial discharge capacity of 1946.5 mAh g^{-1} with an initial C.E. of 57.3%, much higher than that of MoS_2 (1579.2 mAh g^{-1}, 55.4%) shown in Fig. 2.16c.

Importantly, the space among fibers guaranteed the consistent growth of MoS_2 plates on PCNFs and the template effect of fibers banned the overgrowth of MoS_2 into bulk materials, which not only enabled MoS_2 a direct contact with conductive PCNF to shorten the distance of diffusion. Thus ease the electron transfer, but also prohibited the nanoplates from aggregation during the cycling process. Therefore, the composite could keep the high acquaintance of active sites and high specific surface area. High capacities were attained [163]. Also, WS_2 has been introduced as

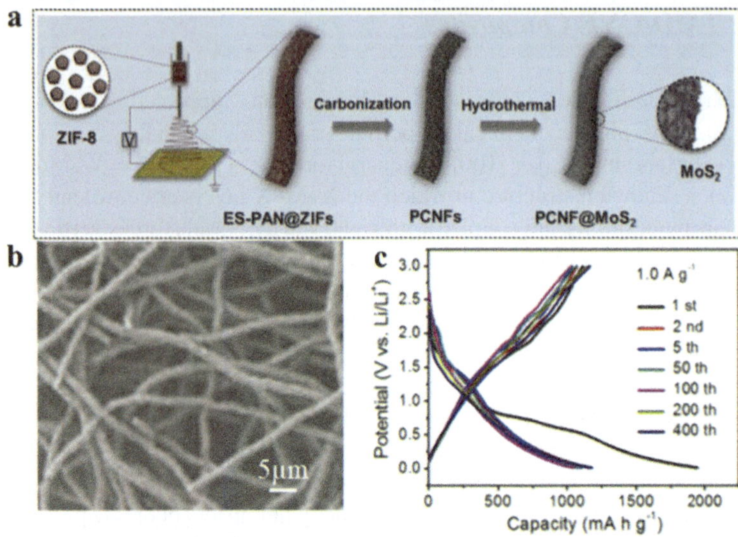

Fig. 2.19 **a** Schematic illustration for the preparation procedure of the PCNF@MoS$_2$ fiber. **b** SEM images of PCNF@MoS$_2$. **c** Potential profiles of PCNF@MoS$_2$ at 1.0 A g^{-1}. Reproduced from Ref. [163]. Copyright 2019, Elsevier

anode material since it has a better intrinsic conductivity than MoS$_2$, which could be a better anode material for LIBs [164]. For example, Yu et al. prepared WS$_2$@NCNFs via a facile electrospinning followed by two-step calcination method under H$_2$/N$_2$ 5%/95%, v/v. As shown in the scheme Fig. 2.19a the atomic structure of single-layered WS$_2$ nanoplates embedded at different angles in the NCNFs. The TEM image proves that single-layered WS$_2$ nanoplates have the thickness of ca. 0.31 nm and the lateral dimension of ca. 5 nm, were uniformly scattered in N-doped CNFs Figs. 2.19b and 2.20.

The battery studies offers high specific capacity (an initial charge capacity of 590.4 mAh g^{-1} at a current density of 0.1 A g^{-1}), a large initial coulombic efficiency (81.7%), excellent capacity retention (437.5 mA h g^{-1} after 200 cycles at a current density of 0.5 A g^{-1}), and great rate capability (see Fig. 2.19c) of 367.1 mAh g^{-1} at a current density of 2 A g^{-1}. These results can be endorsed to many factors as follows: (i) the single layers of WS$_2$ nanoplates can provide numerous interfacial Li-storage sites (ii) the carbon matrix can inhibit the restacking of WS$_2$ single-layers and the aggregation of W nanoparticles after cycling; (iii) the 1-D inter-connected carbon networks can buffer the volume expansion and facilitate charge transfers during lithiation/delithiation; and (iv) N-doping in CNFs can enhance the electrical conductivity [165]. Another, TMD is cobalt selenides (CoSe$_x$) which have

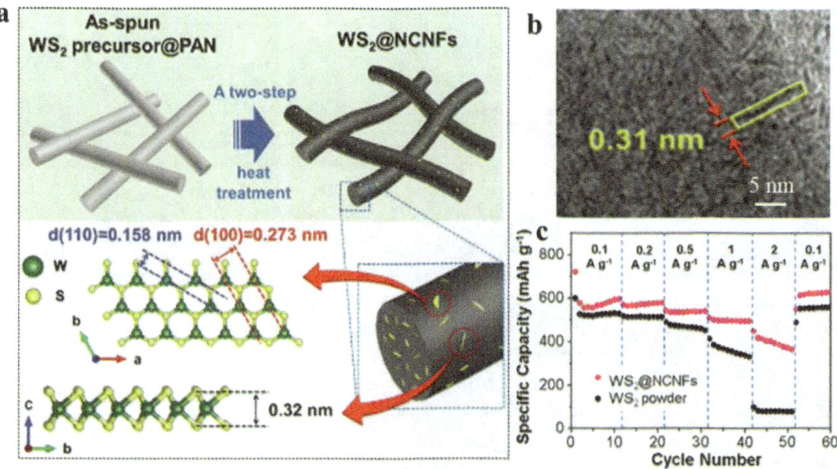

Fig. 2.20 **a** Schematic illustration of the synthetic procedure of WS$_2$@NCNFs. **b** TEM images of WS$_2$@NCNFs. **c** Rate capabilities of WS$_2$@NCNFs. Reproduced from Ref. [165]. Copyright 2015, The Royal Society of Chemistry

been broadly deliberate in the energy storage devices [166, 167]. Recently, one-dimensional CoSe@N-doped carbon nanofibers (CoSe@NC NFs) with electrospinning and annealing, and the moderate amount of selenium powders were directly added in the precursor solution, which could ensure the formation of CoSe. Also, the as-synthesized CoSe@NC NFs material was annealed at different temperatures and the obtained CoSe@NC-550 material exhibit higher capacity and better rate capability in the cycling process.

The TEM and corresponding EDS image of CoSe@NC-550 sample shown in Fig. 2.21a–e. The CoSe@NC-550 obtained the high capacity of 796 mAh g^{-1} at a current density of 1 A g^{-1} for 100 cycles, and the pseudocapacitive contribution is up to 72.8% at the scan rate of 1 mV/s [168]. The cycling life of all samples at different calcination temperatures illustrated in Fig. 2.21f that disclose CoSe@NC-550 has high performance due to the formation of smaller CoSe, high nitrogen content and more defects of the proposed structure. Tin disulfides (SnS$_2$) also has taken particular curiosity as anode for LIBs owing to their high theoretical capacities, and the proposed reaction mechanism is

$$SnS_2 + 4Li \rightarrow Sn + 2Li_2S \tag{2.12}$$

$$Sn + xLi \leftrightarrow Li_x Sn (x \leq 4.4) \tag{2.13}$$

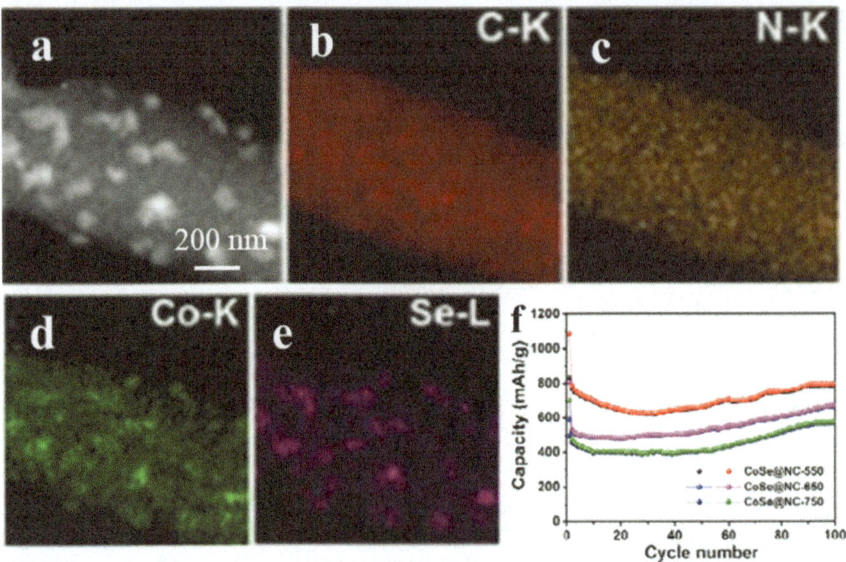

Fig. 2.21 **a** TEM image and **b–e** EDS images of CoSe@NC-550. **f** Cycling performance of CoSe@NC at different calcination temperatures. Reproduced from Ref. [168]. Copyright 2018, Elsevier

The 1st step leads to the 1st irreversible capacity, while the 2nd step provides to a reversible theoretical capacity of 645 mAh g^{-1}, considerably higher than that of the commercial graphite anodes. Yet, fast capacity decay occurred during cycling due to huge volume change. Though many studies investigated to overcome intrinsic properties of SnS$_2$, the electrospun based SnS$_2$ have a particular interest towards high performance. Previously, Kong et al. reported SnS$_2$@graphene nanocable structure via electrospinning method followed by calcination at high temperature. SnS$_2$@graphene nanocable shows around 150 nm in diameter and several micrometers in length. When examining its electrochemical test delivered an outstanding cycling life with a specific capacity of 720 mAh g^{-1} even after 350 cycles at a current density of 0.2 A g^{-1} and over 93.5% capacity retention, but put on display a high-rate capability of 580 mAh g^{-1} at a current rate of 1 A g^{-1}. This behaviour attributes to the great adaptation of Li-ions and unique architecture facilitate the easy access of electrolyte to the active electrode materials [169]. Table 2.2 summarises various type of electrospun based anode materials for LIBs.

Table 2.2 Different types of electrospun based anode materials for LIBs

Electrodes	Precursors	Process	LIB Performance	References
Sb-doped SnO_2 NFs	$SbCl_3 \cdot H_2O$, $SnCl_2 \cdot 2H_2O$/PVP/MeOH	600 °C for 6 h in air	750 mAh g^{-1} after 100 cycles at 100 mA g^{-1}	[170]
Porous SnO_2 NFs	$SnCl_2 \cdot 2H_2O$/PVP/DMF, EtOH	750 °C for 3 h in air	597 mAh g^{-1} after 300 cycles at 160 mA g^{-1}	[171]
Porous NiO–ZnO hybrid	$Ni(NO_3)_2 \cdot 6H_2O$, $Zn(NO_3)_2 \cdot 6H_2O$/PVP/EtOH, Water	500 °C for 3 h in air	949 mAh g^{-1} after 120 cycles at 200 mA g^{-1}	[133]
Fe_2O_3/SnO_2 coaxial NFs	$FeCl_2 \cdot 4H_2O$, $SnCl_2 \cdot 2H_2O$/PVP/DMF	500 °C for 4 h in air	850 mAh g^{-1} after 80 cycles at 100 mA g^{-1}	[172]
$NiFe_2O_4$ NFs	$Ni(ac)_2 \cdot 4H_2O$, $Fe(NO_3)_3 \cdot 9H_2O$/PVP/DMF, EtOH	800 °C for 5 h in air	514 mAh g^{-1} after 60 cycles at 50 mA g^{-1}	[173]
$Zn_{1-x}Mn_xFe_2O_4$ NF	$Zn(ac)_2 \cdot 2H_2O$, $Mn(ac)_2 \cdot 4H_2O$, $Fe(NO_3)_3 \cdot 9H_2O$/PVP/Water, AcOH	450 °C in air	612 mAh g^{-1} after 50 cycles at 60 mA g^{-1}	[174]
$CuFe_2O_4$ NFs	$Cu(ac)_2 \cdot 4H_2O$, $Fe(NO_3)_3 \cdot 9H_2O$/PVP/DMF	800 °C for 5 h in air	572.4 mAh g^{-1} after 60 cycles at 50 mA g^{-1}	[175]
$TiNb_2O_7$ NFs	$Nb(OC_2H_5)_5$, $C_{16}H_{36}O_4Ti$/PVP/AcOH, EtOH	1st: 300 °C for 1 h in H_2/N_2 (H_2, 5%). 2nd 1000 °C for 4 h in air	221 mA h g^{-1} after 100 cycles at 150 mA g^{-1}	[176]
Hollow TiO_2 NFs	$Ti[OCH(CH_3)_2]_4$/PVP, PEO/DMF, AcOH, EtOH	450 °C for 1 h in air	130 mA h g^{-1} after 300 cycles at 100 mA g^{-1}	[177]
Si NPs/porous CNF composite	Si NPs/PAN/DMF	1st: 250 °C for 0.5 h in air, 2nd: 1000 °C for 1 h in Ar	1104 mA h g^{-1} after 100 cycles at 0.5 A g^{-1}	[178]
MoS_2@graphene nanocables	$MoCl_5$, TEOS/PVP/DMF, EtOH	350 °C in air 800 °C for 2 h in H_2/H_2S	1150 mA h g^{-1} after 160 cycles at 500 mA g^{-1}	[179]
SnS_2@graphene nanocables	$SnCl_2$, TEOS/PVP/DMF, EtOH	1st: 350 °C in air 2nd: 400 °C for 2 h in H_2S	720 mA h g^{-1} after 350 cycles at 200 mA g^{-1}	[169]
CNFs	–/PAN/DMF	1st: 280 °C for 2 h in air 2nd: 600, 800 and 1000 °C for 12 h in Ar	460 mA h g^{-1} after 550 cycles at 0.1 A g^{-1}	[180]

References

1. J.M. Tarascon, M. Armand, Issues and challenges facing rechargeable lithium batteries. Nature **414**(6861), 359–367 (2001)
2. M. Winter, R.J. Brodd, *What are Batteries, Fuel Cells, and Supercapacitors?* (ACS Publications, 2004)
3. J.B. Goodenough, K.-S. Park, The Li-ion rechargeable battery: a perspective. J. Am. Chem. Soc. **135**(4), 1167–1176 (2013)
4. V. Etacheri, R. Marom, R. Elazari, G. Salitra, D. Aurbach, Challenges in the development of advanced Li-ion batteries: a review. Energy Environ. Sci. **4**(9), 3243–3262 (2011)
5. H. Wang, LiNi0. 5Mn1. 5O4 cathodes for lithium ion batteries: a review. J. Nanosci. Nanotechnol. **15**(9), 6883–6890 (2015)
6. Y. Kim, K.-S. Park, S.-H. Song, J. Han, J.B. Goodenough, Access to M3+/M2+ redox couples in layered LiMS2 sulfides (M= Ti, V, Cr) as anodes for Li-Ion battery. J. Electrochem. Soc. **156**(8), A703–A708 (2009)
7. K. Mizushima, P. Jones, P. Wiseman, J.B. Goodenough, LixCoO2 (0< x<-1): A new cathode material for batteries of high energy density. Mater. Res. Bull. **15**(6), 783–789 (1980)
8. D. Aurbach, B. Markovsky, G. Salitra, E. Markevich, Y. Talyossef, M. Koltypin, L. Nazar, B. Ellis, D. Kovacheva, Review on electrode–electrolyte solution interactions, related to cathode materials for Li-ion batteries. J. Power Sources **165**(2), 491–499 (2007)
9. G.G. Amatucci, J.M. Tarascon, L.C. Klein, Cobalt dissolution in LiCoO2-based non-aqueous rechargeable batteries. Solid State Ion. Diffus. React. **83**(1–2), 167–173 (1996)
10. C.H. Lu, S.K. Saha, Morphology and electrochemical properties of LiMn2O4 powders derived from the sol-gel route, Mater. Sci. Eng. B-Solid State Mater. Adv. Technol. **79**(3), 247–250 (2001)
11. L.X. Yuan, Z.H. Wang, W.X. Zhang, X.L. Hu, J.T. Chen, Y.H. Huang, J.B. Goodenough, Development and challenges of LiFePO4 cathode material for lithium-ion batteries. Energy Environ. Sci. **4**(2), 269–284 (2011)
12. T. Ohzuku, Y. Makimura, Layered lithium insertion material of LiCo1/3Ni1/3Mn1/3O2 for lithium-ion batteries. Chem. Lett. (7), 642–643 (2001)
13. P. Larsson, R. Ahuja, A. Nyten, J.O. Thomas, An ab initio study of the Li-ion battery cathode material Li2FeSiO4. Electrochem. Commun. **8**(5), 797–800 (2006)
14. A.P. Tang, Q.W. Zhong, G.R. Xu, H.S. Song, Spherical LiCoBO3 particles prepared via a molten salt method for lithium ion batteries. RSC Adv. **6**(87), 84439–84444 (2016)
15. M.S. Whittingham, Lithium batteries and cathode materials. Chem. Rev. **104**(10), 4271–4302 (2004)
16. B. Xu, D.N. Qian, Z.Y. Wang, Y.S.L. Meng, Recent progress in cathode materials research for advanced lithium ion batteries. Mater. Sci. Eng. R-Rep. **73**(5–6), 51–65 (2012)
17. H. Li, Z. Wang, L. Chen, X. Huang, Research on advanced materials for Li-ion batteries. Adv. Mater. **21**(45), 4593–4607 (2009)
18. M.L. Terranova, S. Orlanducci, E. Tamburri, V. Guglielmotti, M. Rossi, Si/C hybrid nanostructures for Li-ion anodes: an overview. J. Power Sources **246**, 167–177 (2014)
19. L.P. Tan, Z. Lu, H.T. Tan, J. Zhu, X. Rui, Q. Yan, H.H. Hng, Germanium nanowires-based carbon composite as anodes for lithium-ion batteries. J. Power Sources **206**, 253–258 (2012)
20. H. Li, Q. Wang, L. Shi, L. Chen, X. Huang, Nanosized SnSb alloy pinning on hard non-graphitic carbon spherules as anode materials for a Li ion battery. Chem. Mater. **14**(1), 103–108 (2002)
21. T.H. Kim, J.S. Park, S.K. Chang, S. Choi, J.H. Ryu, H.K. Song, The current move of lithium ion batteries towards the next phase. Adv. Energy Mater. **2**(7), 860–872 (2012)
22. X. Su, Q. Wu, J. Li, X. Xiao, A. Lott, W. Lu, B.W. Sheldon, J. Wu, Silicon-based nanomaterials for lithium-ion batteries: a review. Adv. Energy Mater. **4**(1), 1300882 (2014)
23. S. Goriparti, E. Miele, F. De Angelis, E. Di Fabrizio, R.P. Zaccaria, C. Capiglia, Review on recent progress of nanostructured anode materials for Li-ion batteries. J. Power Sources **257**, 421–443 (2014)

24. Y. Wu, P. Zhu, X. Zhao, M. Reddy, S. Peng, B. Chowdari, S. Ramakrishna, Highly improved rechargeable stability for lithium/silver vanadium oxide battery induced via electrospinning technique. J. Mater. Chem. A **1**(3), 852–859 (2013)

25. A.S. Hameed, M. Nagarathinam, M. Reddy, B. Chowdari, J.J. Vittal, Synthesis and electrochemical studies of layer-structured metastable α i-LiVOPO 4. J. Mater. Chem. **22**(15), 7206–7213 (2012)

26. X. Zhang, N. Böckenfeld, F. Berkemeier, A. Balducci, Ionic-liquid-assisted synthesis of nanostructured and carbon-coated Li3V2 (PO4) 3 for high-power electrochemical storage devices. Chemsuschem **7**(6), 1710–1718 (2014)

27. D.-W. Han, S.-J. Lim, Y.-I. Kim, S.H. Kang, Y.C. Lee, Y.-M. Kang, Facile lithium ion transport through superionic pathways formed on the surface of Li3V2 (PO4) 3/C for high power Li ion battery. Chem. Mater. **26**(12), 3644–3650 (2014)

28. P. Sun, X. Zhao, R. Chen, T. Chen, L. Ma, Q. Fan, H. Lu, Y. Hu, Z. Tie, Z. Jin, Li 3 V 2 (PO 4) 3 encapsulated flexible nanofabric cathodes for fast charging and long life-cycle lithium-ion batteries. Nanoscale **8**(14), 7408–7415 (2016)

29. X. Li, Y. Chen, J. Zou, X. Zeng, L. Zhou, H. Huang, Stable freestanding Li-ion battery cathodes by in situ conformal coating of conducting polypyrrole on NiS-carbon nanofiber films. J. Power Sources **331**, 360–365 (2016)

30. C. Liu, Y. Zhao, R. Yi, Y. Sun, Y. Li, L. Yang, I. Mitrovic, S. Taylor, P. Chalker, C. Zhao, Alloyed Cu/Si core-shell nanoflowers on the three-dimensional graphene foam as an anode for lithium-ion batteries. Electrochim. Acta **306**, 45–53 (2019)

31. J. Kong, Y. Wei, Silicon nanoparticles confined in thin carbon network: the anode of lithium ion batteries with high performance and easy recyclability. J. Electrochem. Soc. **166**(10), A2013–A2020 (2019)

32. H. Wu, G. Zheng, N. Liu, T.J. Carney, Y. Yang, Y. Cui, Engineering empty space between Si nanoparticles for lithium-ion battery anodes. Nano Lett. **12**(2), 904–909 (2012)

33. Z.-L. Xu, B. Zhang, Z.-Q. Zhou, S. Abouali, M.A. Garakani, J. Huang, J.-Q. Huang, J.-K. Kim, Carbon nanofibers containing Si nanoparticles and graphene-covered Ni for high performance anodes in Li ion batteries. RSC Adv. **4**(43), 22359–22366 (2014)

34. Z.-L. Xu, B. Zhang, J.-K. Kim, Electrospun carbon nanofiber anodes containing monodispersed Si nanoparticles and graphene oxide with exceptional high rate capacities. Nano Energy **6**, 27–35 (2014)

35. E. Qu, T. Chen, Q. Xiao, G. Lei, Z. Li, Freestanding silicon/carbon nanofibers composite membrane as a flexible anode for Li-Ion battery. J. Power Sources **403**, 103–108 (2018)

36. W. Li, Z. Yang, Y. Jiang, Z. Yu, L. Gu, Y. Yu, Crystalline red phosphorus incorporated with porous carbon nanofibers as flexible electrode for high performance lithium-ion batteries. Carbon **78**, 455–462 (2014)

37. D. Li, D. Wang, K. Rui, Z. Ma, L. Xie, J. Liu, Y. Zhang, R. Chen, Y. Yan, H. Lin, Flexible phosphorus doped carbon nanosheets/nanofibers: Electrospun preparation and enhanced Li-storage properties as anodes for lithium ion batteries. J. Power Sources **384**, 27–33 (2018)

38. T. Xu, D. Li, S. Chen, Y. Sun, H. Zhang, Y. Xia, D. Yang, Nanoconfinement of red phosphorus nanoparticles in seaweed-derived hierarchical porous carbonaceous fibers for enhanced lithium ion storage. Chem. Eng. J. **345**, 604–610 (2018)

39. S. Liang, X. Pei, W. Jiang, Z. Xu, W. Wang, K. Teng, C. Wang, H. Fu, X. Zhang, Dual-network red phosphorus@ porous multichannel carbon nanofibers/carbon nanotubes as a stable anode for lithium-ion batteries. Electrochim. Acta **322**, 134696 (2019)

40. N. Chen, L. Chen, W. Li, Z. Xu, X. Qian, Y. Liu, *Brush-like Ni/Carbon Nanofibers/Carbon Nanotubes Multi-layer Network for Freestanding Anode in Lithium Ion Batteries* (Ceramics International, 2019)

41. L. Li, P. Liu, K. Zhu, J. Wang, G. Tai, J. Liu, Flexible and robust N-doped carbon nanofiber film encapsulating uniformly silica nanoparticles: long-life and low-cost electrodes for Li-and Na-Ion batteries. Electrochim. Acta **235**, 79–87 (2017)

42. H. Uchiyama, E. Hosono, I. Honma, H. Zhou, H. Imai, A nanoscale meshed electrode of single-crystalline SnO for lithium-ion rechargeable batteries. Electrochem. Commun. **10**(1), 52–55 (2008)

43. B. Zhang, Y. Yu, Z. Huang, Y.-B. He, D. Jang, W.-S. Yoon, Y.-W. Mai, F. Kang, J.-K. Kim, Exceptional electrochemical performance of freestanding electrospun carbon nanofiber anodes containing ultrafine SnO x particles. Energy Environ. Sci. **5**(12), 9895–9902 (2012)

44. L. Xia, S. Wang, G. Liu, L. Ding, D. Li, H. Wang, S. Qiao, Flexible SnO2/N-doped carbon nanofiber films as integrated electrodes for lithium-ion batteries with superior rate capacity and long cycle life. Small **12**(7), 853–859 (2016)

45. B. Zhang, Z.-L. Xu, J.-K. Kim, In situ grown graphitic carbon/Fe 2 O 3/carbon nanofiber composites for high performance freestanding anodes in Li-ion batteries. RSC Adv. **4**(24), 12298–12301 (2014)

46. B. Joshi, E. Samuel, H.S. Jo, Y.-I. Kim, S. Park, M.T. Swihart, W.Y. Yoon, S.S. Yoon, Carbon nanofibers loaded with carbon nanotubes and iron oxide as flexible freestanding lithium-ion battery anodes. Electrochim. Acta **253**, 479–488 (2017)

47. T. Yang, Z. Chen, H. Zhang, M. Zhang, T. Wang, Multifunctional Cr2O3 quantum nanodots to improve the lithium-ion storage performance of carbon nanofiber networks. Electrochim. Acta **217**, 55–61 (2016)

48. F. Wang, C. Li, J. Zhong, Z. Yang, A flexible core-shell carbon layer MnO nanofiber thin film via host-guest interaction: construction, characterization, and electrochemical performances. Carbon **128**, 277–286 (2018)

49. Y. Liang, N. Li, F. Li, Z. Xu, Y. Hu, M. Jing, K. Teng, X. Yan, J. Shi, Controllable nitrogen doping and specific surface from freestanding TiO2@ carbon nanofibers as anodes for lithium ion battery. Electrochim. Acta **297**, 1063–1070 (2019)

50. S. Gao, D. Zhang, K. Zhu, J.A. Tang, Z. Gao, Y. Wei, G. Chen, Y. Gao, Flexible V2O3/carbon nano-felts as electrode for high performance lithium ion batteries. J. Alloy. Compd. **702**, 13–19 (2017)

51. B.N. Joshi, S. An, H.S. Jo, K.Y. Song, H.G. Park, S. Hwang, S.S. Al-Deyab, W.Y. Yoon, S.S. Yoon, Flexible, freestanding, and binder-free SnO x–ZnO/carbon nanofiber composites for lithium ion battery anodes. ACS Appl. Mater. Interfaces. **8**(14), 9446–9453 (2016)

52. B.N. Joshi, S. An, Y.I. Kim, E.P. Samuel, K.Y. Song, I.W. Seong, S.S. Al-Deyab, M.T. Swihart, W.Y. Yoon, S.S. Yoon, Flexible freestanding Fe2O3-SnOx-carbon nanofiber composites for Li ion battery anodes. J. Alloy. Compd. **700**, 259–266 (2017)

53. B. Joshi, E. Samuel, M.-W. Kim, S. Park, M.T. Swihart, W.Y. Yoon, S.S. Yoon, Atomic-layer-deposited TiO2-SnZnO/carbon nanofiber composite as a highly stable, flexible and freestanding anode material for lithium-ion batteries. Chem. Eng. J. **338**, 72–81 (2018)

54. H. Li, H. Ye, Z. Xu, C. Wang, J. Yin, H. Zhu, Freestanding MoO 2/Mo 2 C imbedded carbon fibers for Li-ion batteries. Phys. Chem. Chem. Phys. **19**(4), 2908–2914 (2017)

55. B. Joshi, E. Samuel, Y.I. Kim, M.-W. Kim, H.S. Jo, M.T. Swihart, W.Y. Yoon, S.S. Yoon, Hierarchically designed ZIF-8-derived Ni@ ZnO/carbon nanofiber freestanding composite for stable Li storage. Chem. Eng. J. **351**, 127–134 (2018)

56. Y. Zhao, J. Wang, C. Ma, L. Cao, Z. Shao, A self-adhesive graphene nanoscroll/nanosheet paper with confined Fe1−xS/Fe3O4 hetero-nanoparticles for high-performance anode material of flexible Li-ion batteries. Chem. Eng. J. **370**, 536–546 (2019)

57. Q. Han, K. Zhu, T. Jin, Z. Yang, Y. Si, Y. Wang, L. Jiao, Electrospun Co/Co3SnC0. 7@ N-CNFs as anode for advanced lithium-ion batteries. J. Alloys Comp. **793**, 646–652 (2019)

58. W. Chang, J.-W. Choi, J.-C. Im, J.K. Lee, Effects of ZnO coating on electrochemical performance and thermal stability of LiCoO2 as cathode material for lithium-ion batteries. J. Power Sources **195**(1), 320–326 (2010)

59. S. Luo, K. Wang, J. Wang, K. Jiang, Q. Li, S. Fan, Binder-free LiCoO2/carbon nanotube cathodes for high-performance lithium ion batteries. Adv. Mater. **24**(17), 2294–2298 (2012)

60. O. Toprakci, H.A. Toprakci, L. Ji, G. Xu, Z. Lin, X. Zhang, Carbon nanotube-loaded electrospun LiFePO4/carbon composite nanofibers as stable and binder-free cathodes for rechargeable lithium-ion batteries. ACS Appl. Mater. Interfaces. **4**(3), 1273–1280 (2012)

61. X. Tian, X. Li, T. Yang, K. Wang, H. Wang, Y. Song, Z. Liu, Q. Guo, Porous worm-like NiMoO4 coaxially decorated electrospun carbon nanofiber as binder-free electrodes for high performance supercapacitors and lithium-ion batteries. Appl. Surf. Sci. **434**, 49–56 (2018)

62. G. Jo, I. Choi, H. Ahn, M.J. Park, Binder-free Ge nanoparticles–carbon hybrids for anode materials of advanced lithium batteries with high capacity and rate capability. Chem. Commun. **48**(33), 3987–3989 (2012)
63. X. Wang, M. Zhang, E. Liu, F. He, C. Shi, C. He, J. Li, N. Zhao, Three-dimensional core-shell Fe2O3@ carbon/carbon cloth as binder-free anode for the high-performance lithium-ion batteries. Appl. Surf. Sci. **390**, 350–356 (2016)
64. J. Qin, Q. Zhang, Z. Cao, X. Li, C. Hu, B. Wei, MnOx/SWCNT macro-films as flexible binder-free anodes for high-performance Li-ion batteries. Nano Energy **2**(5), 733–741 (2013)
65. J. Park, H. Yoo, J. Choi, 3D ant-nest network of α-Fe2O3 on stainless steel for all-in-one anode for Li-ion battery. J. Power Sources **431**, 25–30 (2019)
66. W. Li, Z. Yang, J. Cheng, X. Zhong, L. Gu, Y. Yu, Germanium nanoparticles encapsulated in flexible carbon nanofibers as self-supported electrodes for high performance lithium-ion batteries. Nanoscale **6**(9), 4532–4537 (2014)
67. F. Pantò, Y. Fan, S. Stelitano, E. Fazio, S. Patanè, P. Frontera, P. Antonucci, N. Pinna, S. Santangelo, Electrospun C/GeO2 paper-like electrodes for flexible Li-ion batteries. Int. J. Hydrogen Energy **42**(46), 28102–28112 (2017)
68. F. Pantò, Y. Fan, S. Stelitano, E. Fazio, S. Patanè, P. Frontera, P. Antonucci, N. Pinna, S. Santangelo, Are electrospun fibrous membranes relevant electrode materials for li-ion batteries? The case of the C/Ge/GeO2 composite fibers. Adv. Func. Mater. **28**(23), 1800938 (2018)
69. X. He, Y. Hu, Z. Shen, R. Chen, K. Wu, Z. Cheng, X. Zhang, P. Pan, GeOx ultra-dispersed in microporous carbon nanofibers: a binder-free anode for high performance lithium-ion battery. Electrochim. Acta **246**, 981–989 (2017)
70. J. Wang, W.-L. Song, Z. Wang, L.-Z. Fan, Y. Zhang, Facile fabrication of binder-free metallic tin nanoparticle/carbon nanofiber hybrid electrodes for lithium-ion batteries. Electrochim. Acta **153**, 468–475 (2015)
71. B. Anasori, M.R. Lukatskaya, Y. Gogotsi, 2D metal carbides and nitrides (MXenes) for energy storage. Nat Rev Mater **2**(2), 16098 (2017)
72. B. Mendoza-Sánchez, Y. Gogotsi, Synthesis of two-dimensional materials for capacitive energy storage. Adv. Mater. **28**(29), 6104–6135 (2016)
73. H. Yuan, Y. Jin, J. Lan, Y. Liu, Y. Yu, X. Yang, In situ synthesized SnSe nanorods in a SnO x@ CNF membrane toward high-performance freestanding and binder-free lithium-ion batteries. Inorg. Chem. Front. **5**(4), 932–938 (2018)
74. J.-G. Wang, H. Sun, H. Liu, D. Jin, R. Zhou, B. Wei, Edge-oriented SnS2 nanosheet arrays on carbon paper as advanced binder-free anodes for Li-ion and Na-ion batteries. J. Mater. Chem. A **5**(44), 23115–23122 (2017)
75. R. Chen, X. Xue, Y. Hu, W. Kong, H. Lin, T. Chen, Z. Jin, Intermetallic SnSb nanodots embedded in carbon nanotubes reinforced nanofabric electrodes with high reversibility and rate capability for flexible Li-ion batteries. Nanoscale (2019)
76. M. Lu, K.-F. Wang, In-situ growing graphitic nanotubes on carbon nanofibres as a 3D hierarchical binder-free anode for high-performance Li-ion battery. Micro Nano Lett. **14**(6), 698–700 (2019)
77. L. Tao, Y. Zheng, Y. Zhang, H. Ma, M. Di, Z. Zheng, Liquefied walnut shell-derived carbon nanofibrous mats as highly efficient anode materials for lithium ion batteries. RSC Adv. **7**(43), 27113–27120 (2017)
78. Z.-W. Fu, J. Ma, Q.-Z. Qin, Nanostructured LiCoO2 and LiMn2O4 fibers fabricated by a high frequency electrospinning. Solid State Ion. **176**(17–18), 1635–1640 (2005)
79. S. Zhan, Y. Li, H. Yu, LiCoO2 Hollow nanofibers by Co-electrospinning sol-gel precursor. J. Dispersion Sci. Technol. **29**(5), 702–705 (2008)
80. Y.C. Lee, D.-W. Han, M. Park, M.R. Jo, S.H. Kang, J.K. Lee, Y.-M. Kang, Tailored surface structure of LiFePO4/C nanofibers by phosphidation and their electrochemical superiority for lithium rechargeable batteries. ACS Appl. Mater. Interfaces. **6**(12), 9435–9441 (2014)
81. J. Yao, X. Wang, X. Zhao, J. Wang, H. Zhang, W. Yu, G. Liu, X. Dong, Electrospun Li 2 MnO 3-modified Li 1.2 Ni x Co 0.1 Mn 0.9-x O 2 nanofibers: synthesis and enhanced electrochemical performance for lithium-ion batteries. Electron. Mater. Lett. **12**(6), 804–811 (2016)

82. B. Yue, X. Wang, J. Wang, J. Yao, X. Zhao, H. Zhang, W. Yu, G. Liu, X. Dong, Au-doped Li 1.2 Ni 0.7 Co 0.1 Mn 0.2 O 2 electrospun nanofibers: synthesis and enhanced capacity retention performance for lithium-ion batteries. RSC Adv. **8**(8), 4112–4118 (2018)
83. Y. Gu, D. Chen, X. Jiao, Synthesis and electrochemical properties of nanostructured LiCoO2 fibers as cathode materials for lithium-ion batteries. J. Phys. Chem. B **109**(38), 17901–17906 (2005)
84. K. Bachtin, D. Kramer, V.K. Chakravadhanula, X. Mu, V. Trouillet, M. Kaus, S. Indris, H. Ehrenberg, C. Roth, Activation and degradation of electrospun LiFePO4 battery cathodes. J. Power Sources **396**, 386–394 (2018)
85. C.-S. Kang, C. Kim, J.-E. Kim, J.-H. Lim, J.-T. Son, New observation of morphology of Li [Fe1 − xMnx] PO4 nano-fibers (x= 0, 0.1, 0.3) as a cathode for lithium secondary batteries by electrospinning process. J. Phys. Chem. Solids **74**(4), 536–540 (2013)
86. R. Hongtong, P. Thanwisai, R. Yensano, J. Nash, S. Srilomsak, N. Meethong, Core-shell electrospun and doped LiFePO4/FeS/C composite fibers for Li-ion batteries. J. Alloy. Compd. **804**, 339–347 (2019)
87. R. Hongtong, P. Thanwisai, R. Yensano, J. Nash, S. Srilomsak, N. Meethong, Data on effect of electrospinning conditions on morphology and effect of heat-treatment temperature on the cycle and rate properties of core-shell LiFePO4/FeS/C composite fibers for use as cathodes in Li-ion batteries. Data in Brief **26**, 104364 (2019)
88. D. Xie, G. Cai, Z. Liu, R. Guo, D. Sun, C. Zhang, Y. Wan, J. Peng, H. Jiang, The low temperature electrochemical performances of LiFePO4/C/graphene nanofiber with 3D-bridge network structure. Electrochim. Acta **217**, 62–72 (2016)
89. S. Liu, Y.Z. Long, H.D. Zhang, B. Sun, C.C. Tang, H.L. Li, G.W. Kan, C. Wang, Preparation and electrochemical properties of LiMn2O4 nanofibers via electrospinning for lithium ion batteries. Adv. Mater. Res. Trans. Tech. Publ. 799–802 (2012)
90. S. Jayaraman, V. Aravindan, P.S. Kumar, W.C. Ling, S. Ramakrishna, S. Madhavi, Synthesis of porous LiMn 2 O 4 hollow nanofibers by electrospinning with extraordinary lithium storage properties. Chem. Commun. **49**(59), 6677–6679 (2013)
91. J.-G. Wang, H. Liu, H. Liu, X. Li, D. Nan, F. Kang, Electrospun LiMn1. 5Ni0. 5O4 hollow nanofibers as advanced cathodes for high rate and long cycle life Li-ion batteries. J. Alloys Compd. **729**, 354–359 (2017)
92. C.F. Armer, J.S. Yeoh, X. Li, A. Lowe, Electrospun vanadium-based oxides as electrode materials. J. Power Sources **395**, 414–429 (2018)
93. Y. Liu, D. Guan, G. Gao, X. Liang, W. Sun, K. Zhang, W. Bi, G. Wu, Enhanced electrochemical performance of electrospun V2O5 nanotubes as cathodes for lithium ion batteries. J. Alloy. Compd. **726**, 922–929 (2017)
94. Y.L. Cheah, R. von Hagen, V. Aravindan, R. Fiz, S. Mathur, S. Madhavi, High-rate and elevated temperature performance of electrospun V2O5 nanofibers carbon-coated by plasma enhanced chemical vapour deposition. Nano Energy **2**(1), 57–64 (2013)
95. G.-H. An, D.-Y. Lee, H.-J. Ahn, Carbon-encapsulated hollow porous vanadium-oxide nanofibers for improved lithium storage properties. ACS Appl. Mater. Interfaces. **8**(30), 19466–19474 (2016)
96. Y.L. Cheah, N. Gupta, S.S. Pramana, V. Aravindan, G. Wee, M. Srinivasan, Morphology, structure and electrochemical properties of single phase electrospun vanadium pentoxide nanofibers for lithium ion batteries. J. Power Sources **196**(15), 6465–6472 (2011)
97. Y.L. Cheah, V. Aravindan, S. Madhavi, Improved elevated temperature performance of Al-intercalated V2O5 electrospun nanofibers for lithium-ion batteries. ACS Appl. Mater. Interfaces. **4**(6), 3270–3277 (2012)
98. Y.L. Cheah, V. Aravindan, S. Madhavi, Synthesis and enhanced lithium storage properties of electrospun V2O5 nanofibers in full-cell assembly with a spinel Li4Ti5O12 anode. ACS Appl. Mater. Interfaces. **5**(8), 3475–3480 (2013)
99. X. Miao, H. Ni, X. Sha, T. Wang, J. Fang, G. Yang, Preparation of 0.4 Li2MnO3· 0.6 LiNi1/3Co1/3Mn1/3O2 with tunable morphologies via polyacrylonitrile as a template and applications in lithium-ion batteries. J. Appl. Polym. Sci. **133**(10) (2016)

100. D. Ma, Y. Li, P. Zhang, A.J. Cooper, A.M. Abdelkader, X. Ren, L. Deng, Mesoporous Li1. 2Mn0. 54Ni0. 13Co0. 13O2 nanotubes for high-performance cathodes in Li-ion batteries. J Power Sources **311**, 35–41 (2016)

101. Y. Gu, F. Jian, Hollow LiNi0. 8Co0. 1Mn0. 1O2− MgO coaxial fibers: sol− gel method combined with co-electrospun preparation and electrochemical properties. J. Phys. Chem. C **112**(51), 20176–20180 (2008)

102. E. Hosono, T. Saito, J. Hoshino, Y. Mizuno, M. Okubo, D. Asakura, K. Kagesawa, D. Nishio-Hamane, T. Kudo, H. Zhou, Synthesis of LiNi 0.5 Mn 1.5 O 4 and 0.5 Li 2 MnO 3–0.5 LiNi 1/3 Co 1/3 Mn 1/3 O 2 hollow nanowires by electrospinning. CrystEngComm **15**(14), 2592–2597 (2013)

103. N. Arun, V. Aravindan, S. Jayaraman, N. Shubha, W.C. Ling, S. Ramakrishna, S. Madhavi, Exceptional performance of a high voltage spinel LiNi 0.5 Mn 1.5 O 4 cathode in all one dimensional architectures with an anatase TiO 2 anode by electrospinning. Nanoscale, **6**(15), 8926–8934 (2014)

104. B.-B. Wei, Y.-B. Wu, F.-Y. Yu, Y.-N. Zhou, Preparation and electrochemical properties of carbon-coated LiFePO 4 hollow nanofibers. Int. J. Min. Metall. Mater. **23**(4), 474–480 (2016)

105. R. von Hagen, H. Lorrmann, K.C. Möller, S. Mathur, Electrospun LiFe1− yMnyPO4/C nanofiber composites as self-supporting cathodes in li-ion batteries. Adv. Energy Mater. **2**(5), 553–559 (2012)

106. D. Shao, J. Wang, X. Dong, W. Yu, G. Liu, F. Zhang, L. Wang, Coaxial electrospinning fabrication and electrochemical properties of LiFePO 4/C/Ag composite hollow nanofibers. J. Mater. Sci.: Mater. Electron. **24**(12), 4718–4724 (2013)

107. H.g. Wang, D.l. Ma, Y. Huang, X.b. Zhang, Electrospun V2O5 nanostructures with controllable morphology as high-performance cathode materials for lithium-ion batteries. Chem. Eur. J. **18**(29), 8987–8993 (2012)

108. X. Xia, X. Wang, H. Zhou, X. Niu, L. Xue, X. Zhang, Q. Wei, The effects of electrospinning parameters on coaxial Sn/C nanofibers: morphology and lithium storage performance. Electrochim. Acta **121**, 345–351 (2014)

109. L. Ji, Z. Lin, R. Zhou, Q. Shi, O. Toprakci, A.J. Medford, C.R. Millns, X. Zhang, Formation and electrochemical performance of copper/carbon composite nanofibers. Electrochim. Acta **55**(5), 1605–1611 (2010)

110. B.-O. Jang, S.-H. Park, W.-J. Lee, Electrospun Co–Sn alloy/carbon nanofibers composite anode for lithium ion batteries. J. Alloy. Compd. **574**, 325–330 (2013)

111. X. Xia, Z. Li, H. Zhou, Y. Qiu, C. Zhang, The effect of deep cryogenic treatment on SnSb/C nanofibers anodes for Li-ion battery. Electrochim. Acta **222**, 765–772 (2016)

112. H.-R. Jung, W.-J. Lee, Preparation and characterization of Ni-Sn/carbon nanofibers composite anode for lithium ion battery. J. Electrochem. Soc. **158**(6), A644–A652 (2011)

113. Y.-W. Lee, D.-M. Kim, S.-J. Kim, M.-C. Kim, H.-S. Choe, K.-H. Lee, J.I. Sohn, S.N. Cha, J.M. Kim, K.-W. Park, In situ synthesis and characterization of Ge embedded electrospun carbon nanostructures as high performance anode material for lithium-ion batteries. ACS Appl. Mater. Interfaces. **8**(11), 7022–7029 (2016)

114. S. Li, C. Chen, K. Fu, R. White, C. Zhao, P.D. Bradford, X. Zhang, Nanosized Ge@ CNF, Ge@ C@ CNF and Ge@ CNF@ C composites via chemical vapour deposition method for use in advanced lithium-ion batteries. J. Power Sources **253**, 366–372 (2014)

115. J. Wu, X. Qin, C. Miao, Y.-B. He, G. Liang, D. Zhou, M. Liu, C. Han, B. Li, F. Kang, A honeycomb-cobweb inspired hierarchical core–shell structure design for electrospun silicon/carbon fibers as lithium-ion battery anodes. Carbon **98**, 582–591 (2016)

116. H. Wang, X. Yang, Q. Wu, Q. Zhang, H. Chen, H. Jing, J. Wang, S.-B. Mi, A.L. Rogach, C. Niu, Encapsulating silica/antimony into porous electrospun carbon nanofibers with robust structure stability for high-efficiency lithium storage. ACS Nano **12**(4), 3406–3416 (2018)

117. V.A. Agubra, L. Zuniga, D. De la Garza, L. Gallegos, M. Pokhrel, M. Alcoutlabi, Forcespinning: a new method for the mass production of Sn/C composite nanofiber anodes for lithium ion batteries. Solid State Ion. **286**, 72–82 (2016)

118. Z. Li, J. Zhang, J. Shu, J. Chen, C. Gong, J. Guo, L. Yu, J. Zhang, Carbon nanofibers with highly dispersed tin and tin antimonide nanoparticles: preparation via electrospinning and application as the anode materials for lithium-ion batteries. J. Power Sources **381**, 1–7 (2018)

119. S. Kalluri, K.H. Seng, Z. Guo, H.K. Liu, S.X. Dou, Electrospun lithium metal oxide cathode materials for lithium-ion batteries. RSC Adv. **3**(48), 25576–25601 (2013)

120. S. Peng, G. Jin, L. Li, K. Li, M. Srinivasan, S. Ramakrishna, J. Chen, Multi-functional electrospun nanofibres for advances in tissue regeneration, energy conversion and storage, and water treatment. Chem. Soc. Rev. **45**(5), 1225–1241 (2016)

121. L. Li, S. Peng, J.K.Y. Lee, D. Ji, M. Srinivasan, S. Ramakrishna, Electrospun hollow nanofibers for advanced secondary batteries. Nano Energy **39**, 111–139 (2017)

122. Z. Yang, G. Du, C. Feng, S. Li, Z. Chen, P. Zhang, Z. Guo, X. Yu, G. Chen, S. Huang, Synthesis of uniform polycrystalline tin dioxide nanofibers and electrochemical application in lithium-ion batteries. Electrochim. Acta **55**(19), 5485–5491 (2010)

123. Y.J. Hong, J.W. Yoon, J.H. Lee, Y.C. Kang, A new concept for obtaining SnO2 fiber-in-tube nanostructures with superior electrochemical properties. Chem. Eur. J. **21**(1), 371–376 (2015)

124. M. Fehse, S. Cavaliere, P. Lippens, I. Savych, A. Iadecola, L. Monconduit, D. Jones, J. Roziere, F. Fischer, C. Tessier, Nb-doped TiO2 nanofibers for lithium ion batteries. J. Phys. Chem. C **117**(27), 13827–13835 (2013)

125. J. Sundaramurthy, V. Aravindan, P. Suresh Kumar, S. Madhavi, S. Ramakrishna, Electrospun TiO2−δ nanofibers as insertion anode for li-ion battery applications. J. Phys. Chem. C **118**(30), 16776–16781 (2014)

126. G. Zhang, H. Duan, B. Lu, Z. Xu, Electrospinning directly synthesized metal nanoparticles decorated on both sidewalls of TiO 2 nanotubes and their applications. Nanoscale **5**(13), 5801–5808 (2013)

127. G. Yang, L. Wang, S. Peng, J. Wang, D. Ji, W. Yan, S. Ramakrishna, In situ fabrication of hierarchically branched TiO2 nanostructures: enhanced performance in photocatalytic h2 evolution and li–ion batteries. Small **13**(47), 1702357 (2017)

128. V. Aravindan, P.S. Kumar, J. Sundaramurthy, W.C. Ling, S. Ramakrishna, S. Madhavi, Electrospun NiO nanofibers as high performance anode material for Li-ion batteries. J. Power Sources **227**, 284–290 (2013)

129. X. Yan, X. Tong, J. Wang, C. Gong, M. Zhang, L. Liang, Synthesis of hollow nickel oxide nanotubes by electrospinning with structurally enhanced lithium storage properties. Mater. Lett. **136**, 74–77 (2014)

130. R. Sahay, P. Suresh Kumar, V. Aravindan, J. Sundaramurthy, W. Chui Ling, S.G. Mhaisalkar, S. Ramakrishna, S. Madhavi, High aspect ratio electrospun CuO nanofibers as anode material for lithium-ion batteries with superior cycleability. J. Phys. Chem. C **116**(34), 18087–18092 (2012)

131. J.S. Cho, Y.J. Hong, Y.C. Kang, Electrochemical properties of fiber-in-tube-and filled-structured TiO2 nanofiber anode materials for lithium-ion batteries. Chem. Eur. J. **21**(31), 11082–11087 (2015)

132. Q. Wu, R. Xu, R. Zhao, X. Zhang, W. Li, G. Diao, M. Chen, Tube-in-tube composite nanofibers with high electrochemistry performance in energy storage applications. Energy Storage Mater. **19**, 69–79 (2019)

133. L. Qiao, X. Wang, X. Sun, X. Li, Y. Zheng, D. He, Single electrospun porous NiO–ZnO hybrid nanofibers as anode materials for advanced lithium-ion batteries. Nanoscale **5**(7), 3037–3042 (2013)

134. V. Aravindan, J. Sundaramurthy, E.N. Kumar, P.S. Kumar, W.C. Ling, R. von Hagen, S. Mathur, S. Ramakrishna, S. Madhavi, Does carbon coating really improves the electrochemical performance of electrospun SnO2 anodes? Electrochim. Acta **121**, 109–115 (2014)

135. J. Liang, C. Yuan, H. Li, K. Fan, Z. Wei, H. Sun, J. Ma, Growth of SnO 2 nanoflowers on N-doped carbon nanofibers as anode for Li-and Na-ion batteries. Nano-Micro Lett. **10**(2), 21 (2018)

136. F. Panto, Y. Fan, P. Frontera, S. Stelitano, E. Fazio, S. Patane, M. Marelli, P. Antonucci, F. Neri, N. Pinna, S. Santangelo, Are electrospun carbon/metal oxide composite fibers relevant electrode materials for Li-Ion batteries? J. Electrochem. Soc. **163**(14), A2930–A2937 (2016)
137. J.-C. Kim, G.-H. Lee, D.-W. Kim, Fabrication and electrochemical performance of Sn-Based nanocomposite fibers via electrospinning. Electron. Mater. Lett. **9**(6), 775–777 (2013)
138. J. Kim, J.Y. Kim, D. Pham-Cong, S.Y. Jeong, J. Chang, J.H. Choi, P.V. Braun, C.R. Cho, Individually carbon-coated and electrostatic-force-derived graphene-oxide-wrapped lithium titanium oxide nanofibers as anode material for lithium-ion batteries. Electrochim. Acta **199**, 35–44 (2016)
139. L. Thirugunanam, S. Kaveri, V. Etacheri, S. Ramaprabhu, M. Dutta, V.G. Pol, Electrospun nanoporous TiO2 nanofibers wrapped with reduced graphene oxide for enhanced and rapid lithium-ion storage. Mater. Charact. **131**, 64–71 (2017)
140. Z. Xu, L. Wang, W. Wang, N. Li, C. Chen, C. Li, C. Yang, H. Fu, L. Kuang, Migration behavior, oxidation state of iron and graphitization of carbon nanofibers for enhanced electrochemical performance of composite anodes. Electrochim. Acta **222**, 385–392 (2016)
141. X. Xu, Y. Wan, J. Liu, Y. Chen, L. Li, X. Wang, G. Xue, D. Zhou, Encapsulating iron oxide@ carbon in carbon nanofibers as stable electric conductive network for lithium-ion batteries. Electrochim. Acta **246**, 766–775 (2017)
142. J. He, S. Zhao, Y. Lian, M. Zhou, L. Wang, B. Ding, S. Cui, Graphene-doped carbon/Fe3O4 porous nanofibers with hierarchical band construction as high-performance anodes for lithium-ion batteries. Electrochim. Acta **229**, 306–315 (2017)
143. J. Liang, X. Gao, J. Guo, C. Chen, K. Fan, J. Ma, Electrospun MoO 2@ NC nanofibers with excellent Li+/Na+ storage for dual applications. Sci. China Mater. **61**(1), 30–38 (2018)
144. R. Zhang, X. Dong, L. Peng, W. Kang, H. Li, The enhanced lithium-storage performance for MnO nanoparticles anchored on electrospun nitrogen-doped carbon fibers. Nanomaterials **8**(9), 733 (2018)
145. J. Wang, H. Wang, F. Li, S. Xie, G. Xu, Y. She, M.K. Leung, T. Liu, Oxidizing solid Co into hollow Co 3 O 4 within electrospun (carbon) nanofibers towards enhanced lithium storage performance. J. Mater. Chem. A **7**(7), 3024–3030 (2019)
146. S. Moon, J. Yun, J.Y. Lee, G. Park, S.S. Kim, K.J. Lee, Mass-production of electrospun carbon nanofiber containing SiOx for lithium-ion batteries with enhanced capacity. Macromol. Mater. Eng. **304**(3), 1800564 (2019)
147. M. Mao, F. Yan, C. Cui, J. Ma, M. Zhang, T. Wang, C. Wang, Pipe-wire TiO2–Sn@ carbon nanofibers paper anodes for lithium and sodium ion batteries. Nano Lett. **17**(6), 3830–3836 (2017)
148. X. Lu, P. Wang, K. Liu, C. Niu, H. Wang, Encapsulating nanoparticulate Sb/MoOx into porous carbon nanofibers via electrospinning for efficient lithium storage. Chem. Eng. J. **336**, 701–709 (2018)
149. W. Luo, X. Hu, Y. Sun, Y. Huang, Electrospun porous ZnCo 2 O 4 nanotubes as a high-performance anode material for lithium-ion batteries. J. Mater. Chem. **22**(18), 8916–8921 (2012)
150. L. Li, S. Peng, J. Wang, Y.L. Cheah, P. Teh, Y. Ko, C. Wong, M. Srinivasan, Facile approach to prepare porous CaSnO3 nanotubes via a single spinneret electrospinning technique as anodes for lithium ion batteries. ACS Appl. Mater. Interfaces. **4**(11), 6005–6012 (2012)
151. G. Yang, W. Chang, W. Yan, Fabrication and characterization of NiTiO 3 nanofibers by sol–gel assisted electrospinning. J. Sol-Gel. Sci. Technol. **69**(3), 473–479 (2014)
152. A. Karaphun, S. Kaewmala, N. Meethong, S. Hunpratub, E. Swatsitang, Electrochemical and magnetic properties of electrospun SrTi 1− x Fe x O 3 (x= 0, 0.05 and 0.10) nanofibers for anodes of Li-ion batteries. J. Supercond. Novel Magnet. **31**(6), 1909–1916 (2018)
153. L. Li, S. Peng, Y. Cheah, Y. Ko, P. Teh, G. Wee, C. Wong, M. Srinivasan, Electrospun hierarchical CaCo2O4 nanofibers with excellent lithium storage properties. Chem. Eur. J. **19**(44), 14823–14830 (2013)
154. P.F. Teh, Y. Sharma, Y.W. Ko, S.S. Pramana, M. Srinivasan, Tuning the morphology of ZnMn 2 O 4 lithium ion battery anodes by electrospinning and its effect on electrochemical performance. RSC Adv. **3**(8), 2812–2821 (2013)

155. G. Yang, X. Xu, W. Yan, H. Yang, S. Ding, Facile synthesis of interwoven ZnMn2O4 nanofibers by electrospinning and their performance in Li-ion batteries. Mater. Lett. **128**, 336–339 (2014)
156. Y. Hwangbo, J.-H. Yoo, Y.-I. Lee, Electrospun CoFe2O4 nanofibers as high capacity anode materials for Li-ion batteries. J. Nanosci. Nanotechnol. **17**(10), 7632–7635 (2017)
157. G. Yang, X. Xu, W. Yan, H. Yang, S. Ding, Single-spinneret electrospinning fabrication of CoMn2O4 hollow nanofibers with excellent performance in lithium-ion batteries. Electrochim. Acta **137**, 462–469 (2014)
158. T. Dong, G. Wang, P. Yang, Electrospun NiFe2O4@ C fibers as high-performance anode for lithium-ion batteries. Diam. Relat. Mater. **73**, 210–217 (2017)
159. G. Zhou, C. Wu, Y. Wei, C. Li, Q. Lian, C. Cui, W. Wei, L. Chen, Tufted NiCo2O4 nanoneedles grown on carbon nanofibers with advanced electrochemical property for lithium ion batteries. Electrochim. Acta **222**, 1878–1886 (2016)
160. C. Tan, H. Zhang, Two-dimensional transition metal dichalcogenide nanosheet-based composites. Chem. Soc. Rev. **44**(9), 2713–2731 (2015)
161. X. Huang, C. Tan, Z. Yin, H. Zhang, 25th anniversary article: hybrid nanostructures based on two-dimensional nanomaterials. Adv. Mater. **26**(14), 2185–2204 (2014)
162. C. Zhu, X. Mu, P.A. van Aken, Y. Yu, J. Maier, Single-layered ultrasmall nanoplates of MoS2 embedded in carbon nanofibers with excellent electrochemical performance for lithium and sodium storage. Angew. Chem. Int. Ed. **53**(8), 2152–2156 (2014)
163. C.-L. Zhang, Z.-H. Jiang, B.-R. Lu, J.-T. Liu, F.-H. Cao, H. Li, Z.-L. Yu, S.-H. Yu, MoS2 nanoplates assembled on electrospun polyacrylonitrile-metal organic framework-derived carbon fibers for lithium storage. Nano Energy **61**, 104–110 (2019)
164. S. Zhou, J. Chen, L. Gan, Q. Zhang, Z. Zheng, H. Li, T. Zhai, Scalable production of self-supported WS 2/CNFs by electrospinning as the anode for high-performance lithium-ion batteries. Sci. Bull. **61**(3), 227–235 (2016)
165. S. Yu, J.-W. Jung, I.-D. Kim, Single layers of WS 2 nanoplates embedded in nitrogen-doped carbon nanofibers as anode materials for lithium-ion batteries. Nanoscale **7**(28), 11945–11950 (2015)
166. S.-K. Park, J.K. Kim, Y.C. Kang, Excellent sodium-ion storage performances of CoSe2 nanoparticles embedded within N-doped porous graphitic carbon nanocube/carbon nanotube composite. Chem. Eng. J. **328**, 546–555 (2017)
167. H. Hu, J. Zhang, B. Guan, X.W. Lou, Unusual formation of CoSe@ carbon nanoboxes, which have an inhomogeneous shell, for efficient lithium storage. Angew. Chem. Int. Ed. **55**(33), 9514–9518 (2016)
168. J. Liu, J. Liang, C. Wang, J. Ma, Electrospun CoSe@ N-doped carbon nanofibers with highly capacitive Li storage. J. Energy Chem. **33**, 160–166 (2019)
169. D. Kong, H. He, Q. Song, B. Wang, Q.-H. Yang, L. Zhi, A novel SnS 2@ graphene nanocable network for high-performance lithium storage. RSC Adv. **4**(45), 23372–23376 (2014)
170. Y. Kim, W. Kim, Y.L. Joo, Further improvement of battery performance via charge transfer enhanced by solution-based antimony doping into tin dioxide nanofibers. J. Mater. Chem. A **2**(22), 8323–8327 (2014)
171. S.M. Hwang, Y.-G. Lim, J.-G. Kim, Y.-U. Heo, J.H. Lim, Y. Yamauchi, M.-S. Park, Y.-J. Kim, S.X. Dou, J.H. Kim, A case study on fibrous porous SnO2 anode for robust, high-capacity lithium-ion batteries. Nano energy **10**, 53–62 (2014)
172. W. Xie, S. Li, S. Wang, S. Xue, Z. Liu, X. Jiang, D. He, N-doped amorphous carbon coated Fe3O4/SnO2 coaxial nanofibers as a binder-free self-supported electrode for lithium ion batteries. ACS Appl. Mater. Interfaces. **6**(22), 20334–20339 (2014)
173. L. Luo, R. Cui, K. Liu, H. Qiao, Q. Wei, Electrospun preparation and lithium storage properties of NiFe 2 O 4 nanofibers. Ionics **21**(3), 687–694 (2015)
174. P.F. Teh, S.S. Pramana, Y. Sharma, Y.W. Ko, S. Madhavi, Electrospun Zn1−x Mn x Fe2O4 nanofibers as anodes for lithium-ion batteries and the impact of mixed transition metallic oxides on battery performance. ACS Appl. Mater. Interfaces. **5**(12), 5461–5467 (2013)

175. L. Luo, R. Cui, H. Qiao, K. Chen, Y. Fei, D. Li, Z. Pang, K. Liu, Q. Wei, High lithium electroactivity of electrospun $CuFe_2O_4$ nanofibers as anode material for lithium-ion batteries. Electrochim. Acta **144**, 85–91 (2014)

176. S. Jayaraman, V. Aravindan, P. Suresh Kumar, W. Chui Ling, S. Ramakrishna, S. Madhavi, Exceptional performance of $TiNb_2O_7$ anode in all one-dimensional architecture by electrospinning, ACS Appl. Mater. Interfaces **6**(11), 8660–8666 (2014)

177. X. Zhang, V. Aravindan, P.S. Kumar, H. Liu, J. Sundaramurthy, S. Ramakrishna, S. Madhavi, Synthesis of TiO_2 hollow nanofibers by co-axial electrospinning and its superior lithium storage capability in full-cell assembly with olivine phosphate. Nanoscale **5**(13), 5973–5980 (2013)

178. X. Zhou, L.J. Wan, Y.G. Guo, Electrospun silicon nanoparticle/porous carbon hybrid nanofibers for lithium-ion batteries. Small **9**(16), 2684–2688 (2013)

179. D. Kong, H. He, Q. Song, B. Wang, W. Lv, Q.-H. Yang, L. Zhi, Rational design of MoS_2@ graphene nanocables: towards high performance electrode materials for lithium ion batteries. Energy Environ. Sci. **7**(10), 3320–3325 (2014)

180. Y. Wu, M. Reddy, B. Chowdari, S. Ramakrishna, Long-term cycling studies on electrospun carbon nanofibers as anode material for lithium ion batteries. ACS Appl. Mater. Interfaces **5**(22), 12175–12184 (2013)

Chapter 3
Electrospinning of Nanofibers for Na-Ion Battery

Abstract This chapter describes a comprehensive overview of the preparation of electrospun nanofibers and their applications in advanced secondary Na-ion batteries since NIBs are being considered to be potential alternatives to replace popular LIBs. We next go on to debate the recent research activities involving various electrospun like nanofibers film and porous N-doped fiber in advanced NIBs. The relationship between geometric properties of different fiber structures and their performance also emphasized. Furthermore, we found an impact of the contribution of N to the π-conjugated system of carbon. Likewise, pyridinic-N and pyrrolic-N can make some defects among the CNFs, delivering more open channels and active sites for Na^+ insertion. High energy cathode materials and the interlayer distance have been discussed with appropriate literature. These findings reveal that electrospun nanofibers with unique properties such as large surface area, engineered high porosity, and freestanding characteristics have an immediate effect on the final performance in their applications.

Keywords Binder-free · Freestanding electrode · Morphology effect · Na-ion battery

3.1 Na-Ion Battery Working Principle and Cell Structure

Sodium metal implies the most attractive alternative next to lithium metal, which has the indistinguishable physical and chemical with lithium metal. LIBs and NIBs were explored together during the year of the 1980s [1]. At that period sodium was unnoticed because of the more desirable features of LIBs, particularly energy density. Due to the energy demand, it again got the concern of experts due to cheap raw materials. Furthermore, the SIBs working mechanism resembles the rock-chair type LIBs, which relies on the transport of Li^+ between the electrodes. It consists of two Na inserted anode and cathode materials which are physically separated separator and electronically separated by an electrolyte (in general, electrolyte salts dissolved in aprotic polar solvents), as a pure ionic conductor [2]. A model Na-ion battery working principle could be understanding by the scheme, as shown in Fig. 3.1.

S. Peng and P. R. Ilango, *Electrospinning of Nanofibers for Battery Applications*,
https://doi.org/10.1007/978-981-15-1428-9_3

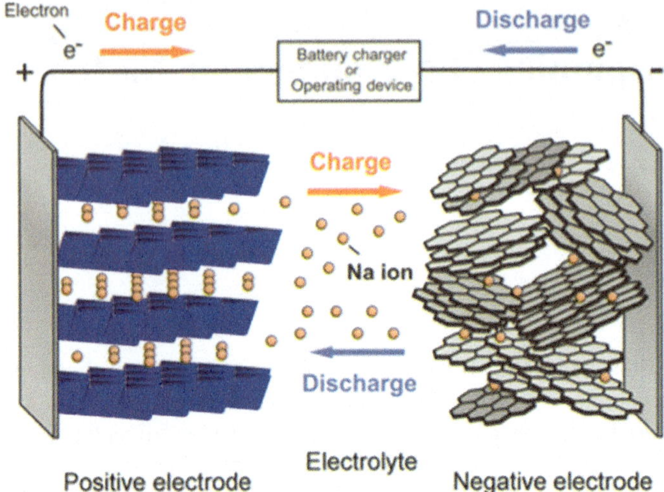

Fig. 3.1 Schematic diagram and cell structure of Na-ion battery. Reproduced from Ref. [3]. Copyright 2014, American Chemical Society

In modern days, ambient-temperature NIBs have elevated much consideration again for grid-level claims since the sustainability benefits of NIBs. Thus, many researchers have put efforts into evolving appropriate electrode materials for NIBs. The battery performance determined by components adopted, and many diverse NIBs for various purposes can be fabricated.

3.2 Perspectives on Material Development

NIBs are vigorously revisited as a possible another to LIBs for large-scale electric storage, due to the natural abundance, low cost, and environmental benignity of Na resources. Nonetheless, the major challenge in advancing NIBs is to find appropriate Na storage electrode materials that are skilled in accommodating more abundant Na^+ ions (1.09 A) with adequate capacity and diffusion kinetics [4]. Carbon-based materials are usually used as anode materials for marketable LIBs. Nevertheless, their low reversible capacity and Na plating performance reduce them inappropriate for NIBs [5, 6]. The development of efficient electrode materials with high capacity as well as outstanding cycle and high rate performance is therefore essential for the commercialization of NIBs.

3.2.1 Cathodes

On behalf of cathode materials, layered metal oxide ($LiMnO_2$) have been used for LIBs. Therefore, it is no surprise that the same $NaMO_2$ compounds have been aimed as Na-intercalated electrodes. Na_xMO_2 materials possess various crystal structures [7]. As the imperative factor of SIBs, cathode materials should be given more devotion, owing to their noteworthy consequence on the electrochemical performance of SIBs in terms of specific energy, cycling life, and specific power. Notably, the Layered materials such as $NaMO_2$ looks more capable since its compounds usually form an O3 structure [8]. Based on the availability of positive electrode materials in Fig. 3.2 for NIBs can be divided into four different types as follows [9]

- Transition metal oxides
- Polyanionic compounds
- Metal hexacyanometalates
- Organic compounds.

There are many kinds of metal oxides will undergo transition-based metal oxides-based cathodes such as single metal oxides like $LiCoO_2$, multi-metal oxides and tunnel oxides. Polyanionic compounds are usually performed by high operating potential and excellent cycling performance. Recently available polyanionic compounds mostly comprise phosphates, pyrophosphates, fluorophosphate and sulphates. Metal hexacyanometalates structure permits reversible de-insertion and insertion of alkali ions due to the open cubic framework and satisfactory interstitial sites. Besides, the available organic cathode materials for SIBs mainly involve

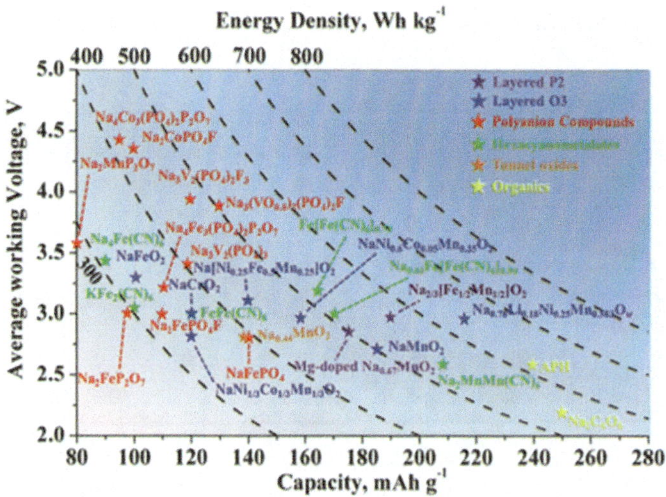

Fig. 3.2 Cathode materials performance data for Na-ion battery. Reproduced from Ref. [2]. Copyright 2018 Wiley–VCH

aromatic carbonyl derivatives, pteridine derivatives, and polymers. In a general viewpoint, the organic compounds are relatively cheap and recyclability [10]

3.2.2   Anodes

Graphite is widely used as negative electrode materials for LIBs owing to its high gravimetric and volumetric capacity due to faster intercalation kinetics. However, a limited amount of Na can be stored in graphite during NIBs performance. Moreover, first-principles calculations specified that it is tough for Na to form the intercalated graphite compounds compared to other alkali metals [11]. Later, so many anode materials have been investigated, and their performance is shown in Fig. 3.3. Explore on negative electrode materials for NIBs can be divided into four different types, identical to those for LIBs as follows [6]

- Carbonaceous materials
- Oxides and phosphates as Na topotactic-insertion materials
- p-block elements (metals, alloys, phosphorus/phosphide)
- Oxides and sulfides showing conversion reaction.

In the recent past, FeS_2, as a conversion-type electrode material, can deliver a high theoretical capacity of 894 mA g^{-1}, which obtainable in nature as pyrite, is inexpensive and eco-friendly. Thus, it remains great challenges to discover actual and simple synthetic procedures for mass-scale manufacture [12].

Moreover, numerous classes of electrode materials have been discovered for SIBs anode and cathode materials to achieve high capacity and long life span. Conversely, most electrode materials explored for SIBs display inferior rate performance and

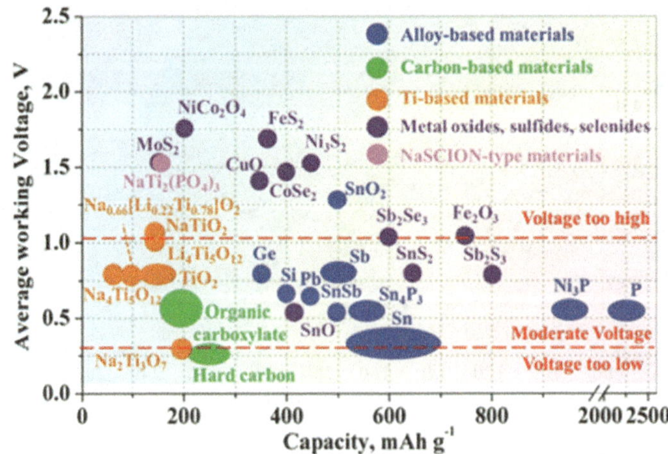

Fig. 3.3   Anode materials performance. Reproduced from Ref. [2]. Copyright 2018 Wiley–VCH

cycle stability owing to the poor electronic conductivity and massive volume changes upon sodiation/de-sodiation. Thus, exploiting rich Na host materials with established structures that concurrently integrate high capacity and long cycle life remains an enormous task.

3.3 Freestanding Electrodes

3.3.1 Freestanding Cathodes

Concerning cathode materials for NIBs, freestanding electrodes have significant attention due to the faster ion diffusion over a high surface area. Jin et al. prepared $NaVPO_4F/C$ nanofibers through electrospinning process and followed by heat treatment. As prepared a $NaVPO_4F$ nanoparticle size is about 6 nm which is embedded in the carbon matrix. As prepared self-standing electrode excellent capacity of 126.3 mAh g^{-1} at 1 °C rate. The TEM images and $NaVPO_4F/C$ cycling performance of self-standing electrodes given in Fig. 3.4a, b [13]. Furthermore, Kretschmer et al. fabricated freestanding $Na_3V_2(PO_4)_3$@CP cathode in which $Na_3V_2(PO_4)_3$ nanoparticles hybridized with carbon paper. As demonstrated in Fig. 3.4c, the nanoparticles are similar to a spherical droplet-like morphology in size range of around 30 nm up to 200 nm, which is the result of the cooperative effect of D-glucose and optimized precursor solution concentration.

The achieved high reversible capacity, exceptional cycle life over 30 000 cycles. Odium ion diffusion coefficient D (cm^2 s^{-1}) can be calculated using the Randles–Sevcik equation

$$I_p = 2.69 * 10^5 n^{3/2} AC\sqrt{D}\sqrt{v} \qquad (3.1)$$

wherein I_p is the peak current (A), A is the footprint area of the electrode, C is the concentration Na$^+$ in a solid (0.0069 mol cm^{-3}), n is the number of electrons. And v is the potential scan rate (V s^{-1}). From CV analysis in Fig. 3.4d, a single pair of defined redox peaks can be observed at all scan rates, which corresponds to the V^{4+}/V^{3+} redox couple and is in good agreement with the charge and discharge voltage plateau [14].

3.3.2 Freestanding Anodes

Flexible Na-ion battery becomes hot research for future power source since its raw material is rich in the earth and very cheap. Nonetheless, the larger ionic radius of Na hinders cycling behaviour which not favour for the practical application. Hence developing freestanding electrodes will be a promising strategy for flexible NIBs

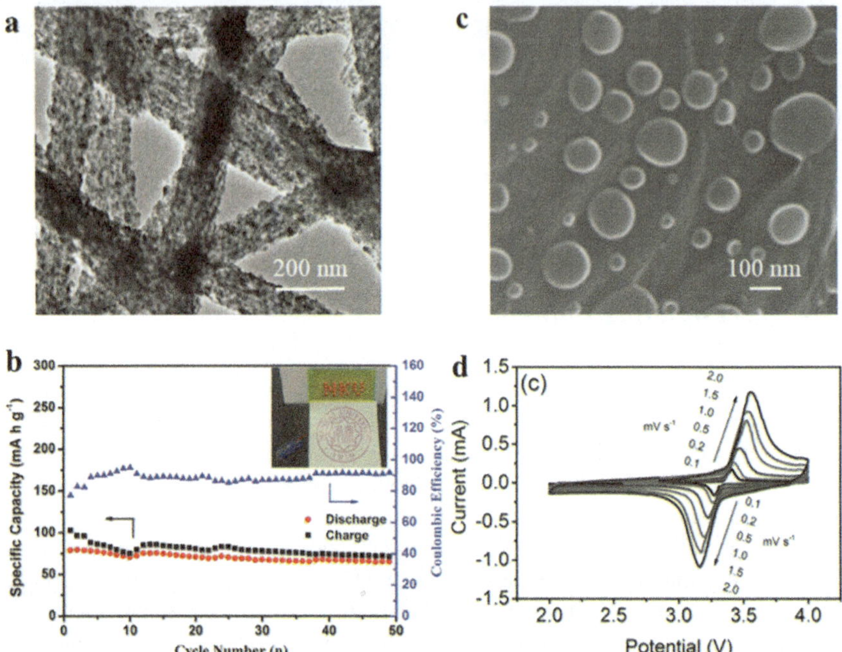

Fig. 3.4 **a** TEM images of NaVPO$_4$F/C nanofibers, **b** cycling performance of NaTi$_2$(PO$_4$)$_3$–NaVPO$_4$F/C full cell (inset is digital picture of a full cell that lights LEDs). Reproduced from Ref. [13]. Copyright 2017 Wiley–VCH, **c** High magnification SEM image of Na$_3$V$_2$(PO$_4$)$_3$ nanoparticles on the carbon fiber surface, **d** cyclic voltammetry (CV) performance of Na 3V$_2$(PO$_4$)$_3$@CP at different scan rates in the voltage range between 2.0 and 4.0 V versus Na$^+$/Na. Reproduced from Ref. [14]. Copyright 2017 Wiley–VCH

[15–17]. Notably, Liu et al. prepared Sn NDs@PNC nanofibers with a flexible free-standing membrane. The SEM image of prepared Sn NDs@PNC nanofibers is illustrated in Fig. 3.5a, with a diameter of about 120 nm, which interlink into a 3D network. As-fabricated Sn NDs@PNC nanofibers delivered a reversible capacity as high as 633 mAh g^{-1} at the current density of 200 mA g^{-1}. The following point would be the main reason for acquiring high performance. First, the doped N can contribute more electrons to the π-conjugated system of carbon, thus enhancing the electronic conductivity of CNFs. Second, the pyridinic-N and pyrrolic-N can make some defects among the CNFs, delivering more open channels and active sites for Na$^+$ insertion [18] see in Fig. 3.5b. Then, Guo et al. synthesized cross-linked carbon nanofiber film, which deals with fast ion diffusion and structural steadiness to bear the frequent impact of Na$^+$ [22]. More excitingly, after thermal treatment, most of the Cu nanoparticles in the CL-CNFs-0.5 film before acid washing are always distributed around the cross-linked points shown in Fig. 3.5c. The CL-CNFs-0.5 film also exhibits superior rate performances than the CNF film (Fig. 3.6d). At 100, 200, 500, 1000 and 2000 mAh g^{-1}, the reversible capacities of the CL-CNFs-0.5 film can

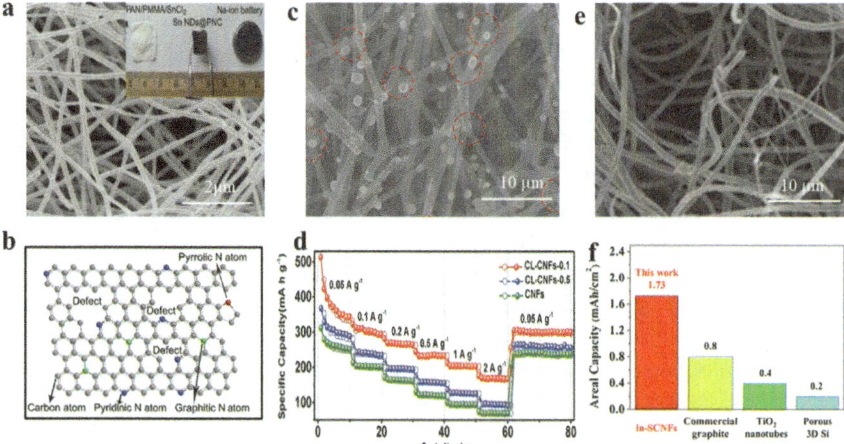

Fig. 3.5 **a** SEM image of Sn NDs@PNC nanofibers. **b** Schematic configurations of the pyridinic-N, pyrrolic-N, and graphitic-N in Sn NDs@PNC nanofibers. Reproduced from Ref. [18] Copyright 2014 Wiley–VCH. **c** SEM image of CL-CNFs-0.5 film before acid washing with clear Cu nanoparticles around cross-linked points. **d** Rate performances of samples CNFs, CL-CNFs-0.5 and CL-CNFs-0.1. Reproduced from Ref. [19]. Copyright 2017, The Royal Society of Chemistry. **e** SEM image of in-SCNFs film after thermal treatment. **f** Areal capacity comparison of the in-SCNFs film. Reproduced from Ref. [20]. Copyright 2017, Elsevier

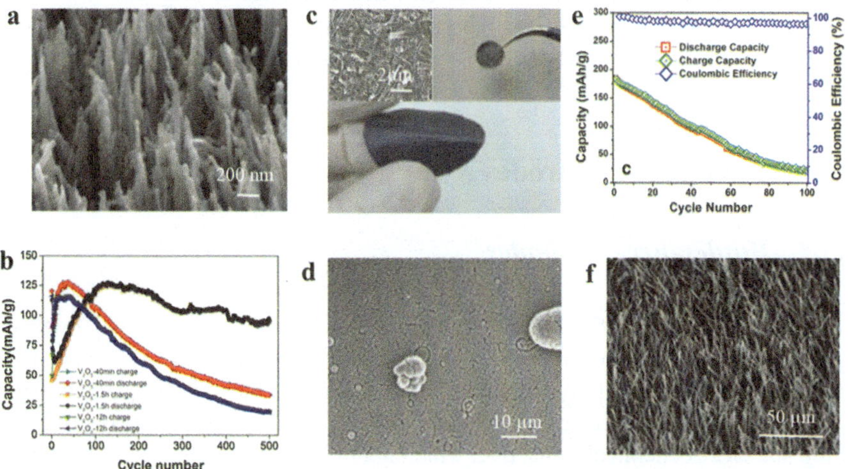

Fig. 3.6 **a** Side-view SEM images of V_2O_5 array on Ti foil at the reaction time of 1.5 h. **b** Cycling of V_2O_5 at a current of 0.25 A g^{-1} at different reaction time. Reproduced from Ref. [25]. Copyright 2015, Elsevier. **c** Digital photograph of freestanding film by $V_2O_5 \cdot 0.34H_2O$ nanoribbons and MWCNT inset SEM images of freestanding film. Reproduced from Ref. [26]. Copyright 2017, Elsevier. **d** Field emission scanning electron micrographs of a-VO_x films deposited at 30 Pa. **e** Na-ion battery cycling performance of a-VO_x-30 Pa films at 0.1 °C. Reproduced from Ref. [31] 2018 Open access. **f** SEM images for a TiS_{2x} vertically oriented nanobelt arrays. Reproduced from Ref. [32]. Copyright 2018, The Royal Society of Chemistry

reach up to 240, 198, 159, 128 and 96 mAh g^{-1}, respectively see Fig. 3.5d [19]. Similarly, Li et al. reported SiO/C freestanding electrode for the first time, which has significant cycling performance for NIBs.

The in-SCNFs become curly and flexible after carbonization, and the finer nanoparticles are uniformly capsulated in carbon nanofibers presented in Fig. 3.5e. It can be seen clearly that the areal capacity Fig. 3.5f of in-SCNFs is 1.73 mAh/cm^2, which is also higher than those of the common-used anodes, such as commercial graphite (0.8 mAh/cm^2), TiO$_2$ nanotubes (0.4 mAh/cm^2) and porous 3D Si (0.2 mAh/cm^2). These results disclose that the in-SCNFs film not only displays good stability and gravimetric capacity but also owns the good volumetric and areal specific capacities [20]. Not long ago, Liu et al. introduced red phosphorus embedded nanofibers as a freestanding electrode for NIBs which possess the superb reversible capacity of 1308 mAh g^{-1} at 200 mA g^{-1} and is one of the best among all phosphorus-based anodes described up to date for NIBs [16]. Also, they proposed that Fe$_2$O$_3$ @C self-standing electrode had 98.3% capacity retention over 1000 cycles when it used as an anode for NIBs. Moreover subsequent cathodic scan, the reduction peak obtained at ~0.70 V is credited to the intrinsic conversion of Fe$_2$O$_3$ with Na$^+$ insertion expressed as

$$Fe_2O_3 + 6Na^+ + 6e^- \rightarrow 2Fe + 3Na_2O \qquad (3.2)$$

For the anodic scan, the two oxidation peaks observed at 0.75 and 1.4 V are endorsed to a two-step Na$^+$ extraction process involving expressed as Fe^0 $\rightarrow$ Fe^{2+} and Fe^{2+} $\rightarrow$ Fe^{3+}, respectively [21]

3.4 Binder-Free Electrodes

3.4.1 Binder-Free Cathodes

The regular fabrication of electrodes typically depends on the slurry coating method, by which viscous slurry of the active materials, conductive carbon and nonconductive binders. In general, polymers are used as binders to blend of active materials, conductive agents, and current collectors. Undesirably, binders are normally insulating and electrochemically inactive, which not only decline the electrical conductivity but also creates side effects with the electrolyte species [22, 23]. Additionally, the binder's increases the weight and volume of the electrode, resulting from the lower energy density of the batteries. To rectify this issue, researchers focused on binder-free electrodes and increased the energy density of the batteries [24]. As of Na-ion batteries, few kinds of literature have been published recently as binder-free cathodes. Among them, V$_2$O$_5$ material has focused extensively as cathode materials due to high energy density and larger interlayer distance. For example, Wang et al. reported V$_2$O$_5$ array on titanium foil via easy hydrothermal method. Moreover, they introduced this protocol for the first time as a binder-free cathode for rechargeable Na-ion batteries and extensively studied that the reaction time plays an essential character in the formation of array type morphology of V_2O_5 shown in Fig. 3.6a, a normal crystallization self-assembly process via Ostwald ripening mechanism [25]. The javelin-like structure provides excellent Na storage performance, as presented in Fig. 3.6b. On the other hand, Sun et al. prepared V_2O_5 nH_2O nanoribbons via a hydrothermal process. Different thickness of ribbons such as 20, 200 and 1 nm thick film was obtained. The V_2O_5 cathode was fabricated by the intertwining nanoribbons with multi-walled carbon nanotubes shown in Fig. 3.6c, resulting in good reversible capacity and cycling stability. Furthermore, they proposed the reaction mechanism when Na-ions intercalate into the V_2O_5 structure [26]

$$V_2O_5 \cdot 0.34H_2O + Na^+ + e^- \rightarrow NaV_2O_5 \cdot 0.34H_2O \tag{3.3}$$

$$NaV_2O_5 \cdot 0.34H_2O + Na^+ + e^- \rightarrow Na_2V_2O_5 \cdot 0.34H_2O \tag{3.4}$$

Similarly, Li et al. attempted $V_2O_5 \cdot H_2O$ films with different V^{4+} contents on a stainless substrate. In this study, sintering plays a crucial role to form V^{4+} in $V_2O_5 \cdot xH_2O$ films. With More V^{4+} defect enhances the Na-storage performances with the initial discharge capacity of 198 mAh g^{-1} and higher capacity retention of 99% after 100 cycles at 200 mA g^{-1} [27]. Kinds of literature recommended that amorphous V_2O_5 can perform better than crystalline structures owing to better space host Na$^+$ ions in the disordered interlayer lattices and the supplementary pseudocapacitive assistances [28, 29]. For instance, Liu et al. conveyed that amorphous V_2O_5 provide a higher capacity of ~250 mAh g^{-1} than the crystalline V_2O_5 at the low rates (40–640 mA g^{-1}), but it was weirdly opposite at high current densities (>6400 mA g^{-1}) [30]. Afterwards, Petnikota et al. proposed an amorphous VO_x thin film which has 650 nm thick on SS substrate. When implementing as a cathode for SIBs, it delivers capacities as high as 164 mAh g^{-1} at 0.1 °C current rate, respectively for the film at 30 Pa [31]. The corresponding SEM images and cycling performance demonstrated in Fig. 3.6d, e. Most recently, Hawkins et al. fabricated Titanium (IV) sulfide (TiS$_2$) as a binder-free electrode for LIBs and NIBs. The fabricated TiS$_2$ nanobelts are highly anisotropic and are vertically grown directly on the current collector to yield a spatially controlled array shown in Fig. 3.6f. It is also noted that the Na-ion battery is comprising our vertically oriented TiS$_2$ nanobelts array exhibit a discharge capacity of 217 mAh g^{-1}. Additionally, they proposed the formation mechanism from TiS$_3$ to TiS$_{2-x}$ as follows

$$2Ti + 6s \rightarrow 2TiS_3 \tag{3.5}$$

$$2TiS_3 \rightarrow 2TiS_{2-x} \tag{3.6}$$

During the desulfurization of TiS_3, nucleation seeds including TiS_2 facets are shaped on the surface of the TiS_3 nanobelts. They are circulated inward at a rate that is powerfully reliant on the reaction time and temperature [32].

3.4.2 Binder-Free Anodes

The employment of polymeric binders will inevitably shrink the electrical conductivity and reduce the availability of active materials that leads to sluggish kinetics during the electrochemical process. There is much remarkable research work done on anode material for Na-ion battery [33–35]. In recent years, a new approach to hypothesis additive-free anodes with acceptable electric conductivity and integrity has been used for amplifying the storage characteristics of anode materials. For example, Yuan et al. prepared CuO nanorod arrays as a binder-free electrode by the facile in-situ engraving method. The FE-SEM images of the prepared electrode have several micrometres in length shown in Fig. 3.7a. Furthermore, this structure would

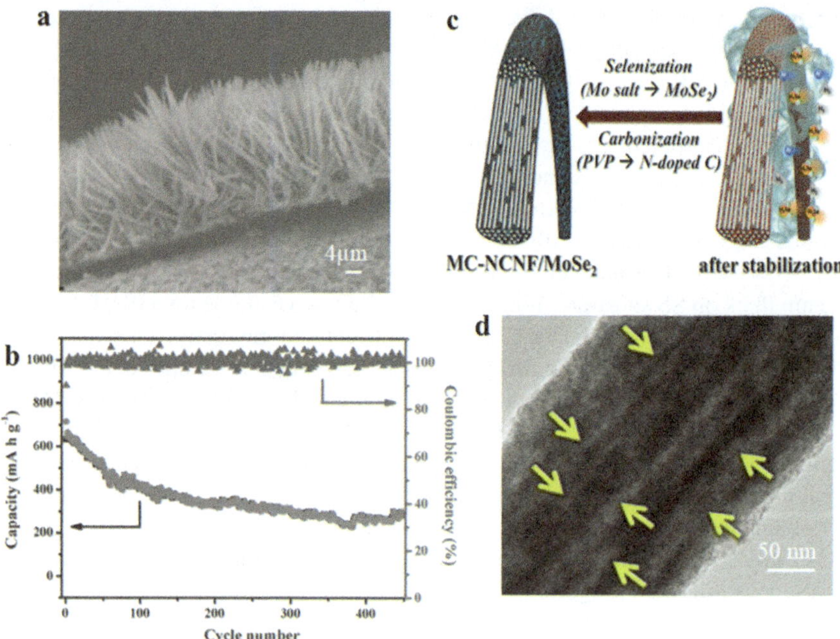

Fig. 3.7 **a** FE-SEM image of CuO nanorod array. **b** Cycling performance of CuO nanorod arrays at 200 mAg^{-1}. Reproduced from Ref. [36]. Copyright 2013 Wiley–VCH. **c** Formation mechanism of MC-NCNF/MoSe$_2$. **d** HR-TEM image of MC-NCNF/MoSe$_2$. Reproduced from Ref. [37]. Copyright 2016, Elsevier

enable faster electron transport resulting in the high capacity of over 640 mAh g^{-1} even at 200 mA g^{-1} [36] presented in Fig. 3.7b.

Later, Yin et al. proposed a full cell in which the Bi_2O_3/C used as an anode, whereas $Na_3V_2(PO_4)_3$ used as a cathode. The Bi_2O_3/C anode has been established to attain a six-electron reaction, and the $Na_3V_2(PO_4)_3$ cathode holds a theoretical 2Na extraction capability. The full cell reaction can be expressed as follows [38]

$$3Na_3V_2(PO_4)_3 + Bi_2O_3 + 6Na^+ \overset{6e^-}{\rightleftharpoons} 3NaV_2(PO_4)_3 + 2Na_3Bi + 3Na_2O \quad (3.7)$$

Differently, Xia et al. reported Nickel sulfide (Ni_2S_2) as a binder-free electrode and studied their Na-storage behaviour. The result concludes that the carbon (graphene and cracked carbon) supported Ni_2S_2 binder-free electrode has advantages of higher electronic conductivity and buffer the volume extension [39, 40]. Similarly, Yin et al. fabricated $CoSe_2/CNFs$, which exhibits superb cycling stability of 430 mAh g^{-1} at 200 mA g^{-1} [41]. Furthermore, Liu et al. reported that graphene monolayers or bilayers highly scattered in CNFs hold a high reversible capacity of 432.3 mAh g^{-1} at 100 mA g^{-1} [42]. Another interest is the binder-free approach of MC-NCNF/$MoSe_2$ in which multi-channel-contained N-doped CNF consisted of few-layered $MoSe_2$ nanosheets. The diethylenetriamine was introduced as a pore generator in the electrospinning process for the first time. Besides, it is played a crucial role in producing multi-channels in the structure by phase-separation from the Mo salt and following volatilization without any additional process [37] shown in Fig. 3.7c and their HR-TEM images shown in Fig. 3.7d.

3.5 Other Composites for Cathode

As of now, many alternative cathode materials have been discussed, such as phosphates [43], fluorophosphate [44] and layered metal oxides [45]. However, the practical capacity and rate capability are still unsatisfied. Electrospinning process is a preferable technique to produce one-dimensional nanofibers which is enhanced the overall performance of cathode. *X. Miao* et al. reported V_2MoO_8 nanofiber cathode for the first time. Figure 3.8a represents the SEM image of VMO product. Its morphology presents broken and short wedged-shaped cuboid, 300–500 nm (thickness) $\times$ 4–6 μm (length). The corresponding initial 2 discharge/charge profile shown in Fig. 3.8b with the different molar ratio of electrolytes. The Addition LiCl salt into electrolyte not only significantly enhances the cycle capacity and voltage plateau to ~ 1.5 V, but also improves the rate performance and cycle stability [45]. Later, Niu et al. proposed NFPO@C@rGO composite in which graphene wrapped $Na_{6.24}Fe_{4.88}(P_2O_7)_4$. This composite has a good discharge capacity of 99 mAh g^{-1} at a current density of 40 mA g^{-1} after 320 cycles, which is 1.6 times higher than that of the pristine $Na_{6.24}Fe_{4.88}(P_2O_7)_4$ (NFPO@C) composite. It concludes that the unique spinning vein fibre-based porous structure granting a decent close contact between

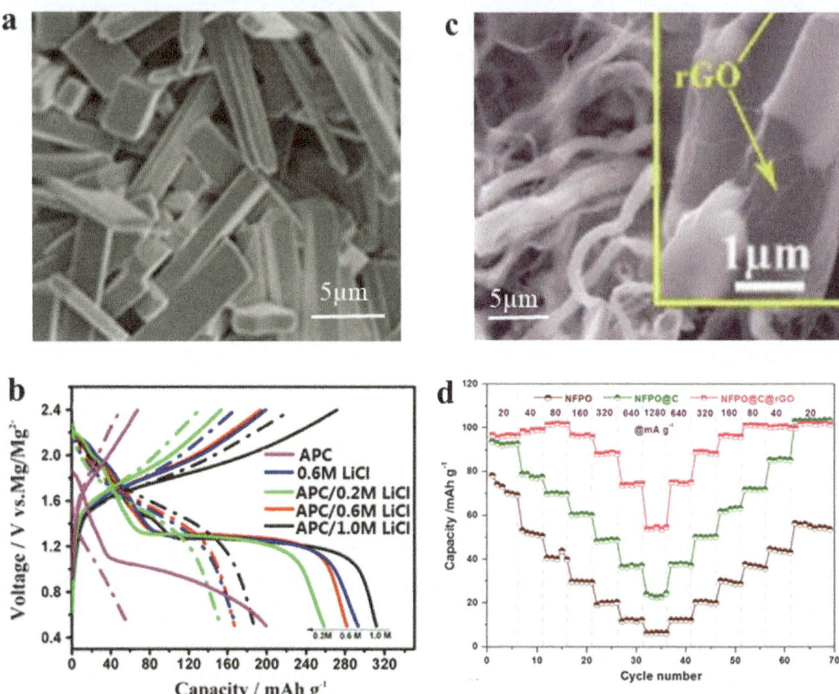

Fig. 3.8 **a** SEM images of VMO obtained by calcination at 630 °C for 6 h. **b** The initial 2 cycle profiles (solid line: the 1st cycle). Reproduced from Ref. [45]. Copyright 2017, Elsevier. **c** FE-SEM images of NFPO@C@rGO cathode. **d** Rate performance of as-prepared cathode fibers. Reproduced from Ref. [46]. Copyright 2017 the Owner Societies

NFPO@C and graphene for high electronic conductivity, fast ionic transport, a large reaction surface and a strong solid structure stopping breakdown during cycling, thus attaining a high rate discharge performance and high cycling stability [46] shown in Fig. 3.8c, d.

3.6 Other Composites for Anodes

3.6.1 Metal/CNF Composites

One dimensional carbon nanofibers with metal-based composites are expected to be stable electrode materials, due to their identical structure, adapted electronic and ionic diffusion, and solid tolerance to stress alteration [47]. Specifically, Sb and Sn are the most attractive anode active material for NIBs since they exposed high reversible capacity [48–50]. In earlier work, Wu et al. reported Sb-C nanofiber which could deliver a large reversible capacity of 631 mA hg^{-1} at C/15 rate. Herein, the total Sb content in the composite was calculated to be 38% based on the following equation [51]

$$Sb(wt\%) = 100 * \frac{Molecular\,weight\,of\,Sb}{Molecular\,weight\,of\,Sb_2O_4} * \frac{Final\,weight\,of\,Sb_2O_4}{Intial\,weight\,of\,Sb - C\,nanofibers}$$
(3.8)

Subsequently, Zhu et al. prepared Sb/C nanofibers in which ~30 nm Sb nanoparticles encapsulated interconnecting 400 nm carbon fibers. The carbonized fibers maintained their fibrous morphology with no obvious diameter change shown in Fig. 3.9a.

The SbNP@C electrode showed an initial total capacity of 422 mAh/g$_{electrode}$ and retained 350 mAh/g$_{electrode}$ over 300 cycles under 100 mA/g$_{Sb}$ [52]. Excitingly, Zhu et al. synthesized core–shell nanostructure of Sb@PCFs, which delivers good rate performance [53] at different current densities provides in Fig. 3.9b. By the way, Sha et al. proposed Sn@NCNFs and it does show excellent high-rate cycling performance and can maintain a capacity of up to 390 mAh g^{-1} even at an extremely high rate of 1 °C for over 1000 cycles. Figure 3.9c displays the first two discharge–charge

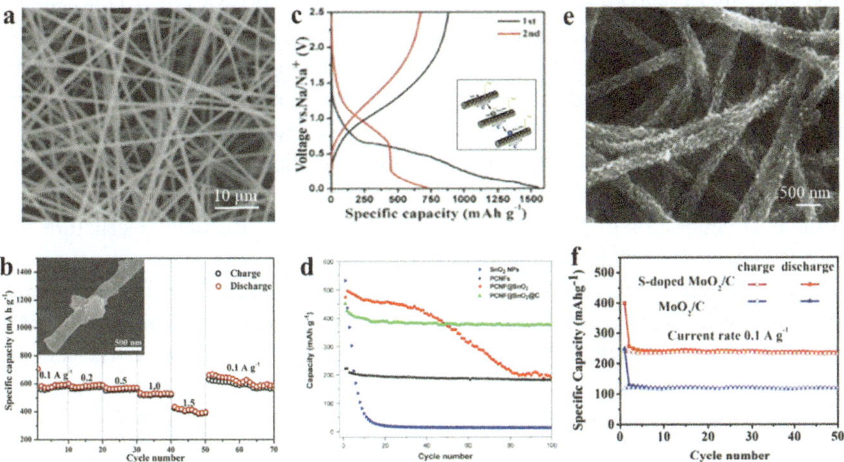

Fig. 3.9 **a** SEM images of SbNP@C nanofibers. Reproduced from Ref. [52]. Copyright 2013, American Chemical Society. **b** Rate performance of Sb@PCFs under 0.1, 0.2, 0.5, 1.0 and 1.5 A g^{-1}, (inset in **c** shows SEM of a 1D fibrous structure after cycling performance). Reproduced from Ref. [53]. Copyright 2017, Elsevier. **c** The first two discharge–charge profiles between 0 and 2.5 V versus Na$^+$/Na at a current rate of 0.1 °C. Reproduced from Ref. [54]. Copyright 2017, The Royal Society of Chemistry. **d** Cycling performance SnO$_2$ nanoparticles, PCNFs, PCNF@SnO$_2$ composite, and PCNF@SnO$_2$@C composite. Reproduced from Ref. [60]. Copyright 2015, American Chemical Society. **e** SEM image of the TiO$_2$/C nanofibers Reproduced from Ref. [61]. Copyright 2016, American Chemical Society. **f** Cycling performances of S-doped MoO$_2$/C and MoO$_2$/C electrodes a 0.1 A g^{-1}. Reproduced from Ref. [62]. Copyright 2018, Elsevier

profiles between 0 V and 2.5 V versus Na$^+$/Na at a current rate of 0.1 °C [54]. On the other hand Bismuth (Bi) based fibers also prepared and reported that Bi/C provides a structurally stable host for ion intercalation and de-intercalation of Na$^+$ [55].

3.6.2 *Metal Oxide/CNF Composites*

Another interesting anode material is a metal oxide with carbon nanofibers. In general, metal oxides such as MoO_2 [56], TiO_2 [57], SnO_2 [58], and FeO_x [59], have high performance for NIBs. Still, they have sluggish cyclic life and rate capability due to their low electronic conductivities. Incorporation of these metal oxides onto the electrospun CNFs have been extensively studied in order to acquire improved Na-storage performance. For example, Dirican et al. proposed PCNF@SnO_2@C in which SnO_2 has been covered by carbon and CNF. Electrochemical results revealed that PCNF@SnO_2@C anode had excellent performance, including high-capacity of 374 mAh g^{-1} with a high coulombic efficiency of 98.9% after 100th cycle [60] shown in Fig. 3.9d. Afterwards, Xiaong et al. were developed TiO_2/C nanofibers, where anatase TiO_2 nanocrystals with a diameter of ~12 nm were densely embedded in the conductive carbon fibers displayed in Fig. 3.9e. TiO_2/C anode shows a high redox capacity of ~302.4 mAh g^{-1} at a high current of 2000 mA g^{-1} [61]. Recently, Li et al. prepared S-doped MoO_2/C nanofibers and yields a high reversible capacity at 0.1 A g^{-1} after 50 cycles of 269 mAh g^{-1} [62] demonstrated in Fig. 3.9f. Despite, very recently, Nie et al. produced TiO_2–Sn/C composite and result reveals that improved specific capacity. Moreover, TiO_2–Sn/C nanofibers reveal an improved diffusion coefficient of Na$^+$ due to a small amount of Sn nanoparticles integration [63]. Meanwhile. binary and ternary metal oxide nanofibers also considered promising anodes for the next generation of NIBs batteries on account of the distinctive structure and rapid electron transport during the electrochemical process [64, 65].

In another study, TiO_2@CNFs composites were introduced at different thermal temperatures to treat one-dimensional electrospun composites in which TiO_2 nanoparticles were distributed in a continuous carbon nanofiber matrix. The scheme indicates the preparation of electrospun nanofibers shown in Fig. 3.10a. The nanofiber precursor was prepared by blending $Ti(OC_4H_9)_4$ dropwise to a 5% PVP solution in C_2H_5OH and CH_3COOH (10:1 by weight) finally treated 1st at 280, and then calcined at 450 °C, 550 °C, 650 °C, and 750 °C, respectively, for 4 h with a heating rate of 2 °C/min. During the process, $Ti(OC_4H_9)_4$ was hydrolyzed to form TiO_2 while PVP was transferred to carbon. The mechanism of TiO_2 formation could be described as the following

$$Ti(OC_4H_9)_4 + 2H_2O \rightarrow TiO_2 + 4C_4H_9OH \tag{3.9}$$

The fibrous structure of the composites was confirmed by TEM images, in which TiO_2 nanoparticles were uniformly and densely distributed provided in Fig. 3.10b. The capacities of TiO_2@CNF composites prepared at 450, 550, 650, and 750 °C

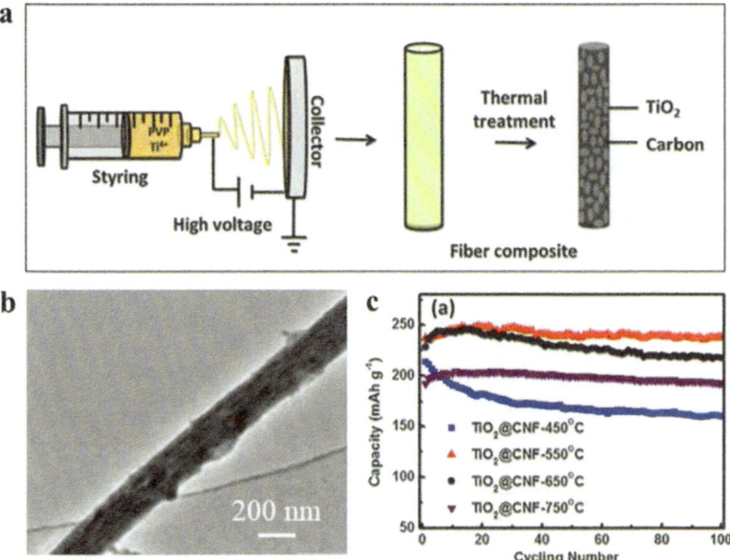

Fig. 3.10 **a** Scheme of the fabrication processes of TiO_2@CNFs composites. **b** TEM images of TiO_2@CNF-550 °C. **c** Cycling performance of TiO_2@CNF composites. Reproduced from Ref. [66]. Copyright 2015, Elsevier

remained at 159.9, 238.1, 217.5, and 193.1 mAh g^{-1} after 100 cycles and their initial coulombic efficiencies were 53.3, 68.2, 59.2, and 51%, respectively Fig. 3.10c. Form this observation; it is suggested that the TiO_2@CNF-550°, with anatase TiO_2 structure, exhibited the most exceptional cycling ability [66].

3.6.3 2D/CNF Composites

Recently, two dimensional material (2D), for example, transition metal chalcogenide materials including metal selenides and sulfides have been considered as anode materials for NIBs. TMD materials, along with carbon nanofibers help to improve the structural stability and Na-storage capacity [37, 67].

In earlier, Ryu et al. synthesised heterogeneous WS_x/WO_3 thorn-bush nanofiber as anodes for NIBs. As shown in Fig. 3.11a, the TEM image reveals nanothorns along the vertical direction of the NFs with average length and diameter of the thorns are 100 and 10 nm, respectively. As-prepared the heterogeneous thorn-bush nanofiber electrodes deliver a high second discharge capacity of 791 mAh g^{-1} for 100 cycles than pristine WS_x nanofiber represented in Fig. 3.11b. This kind of hierarchical design could reduce the sulphur dissolution during the cycling resulting in improved NIB performances [68]. Furthermore, Cho et al. proposed $NiSe_2$-rGO-C composite for the first time in which the rGO wrapped $NiSe_2$/C. These composite nanofibers

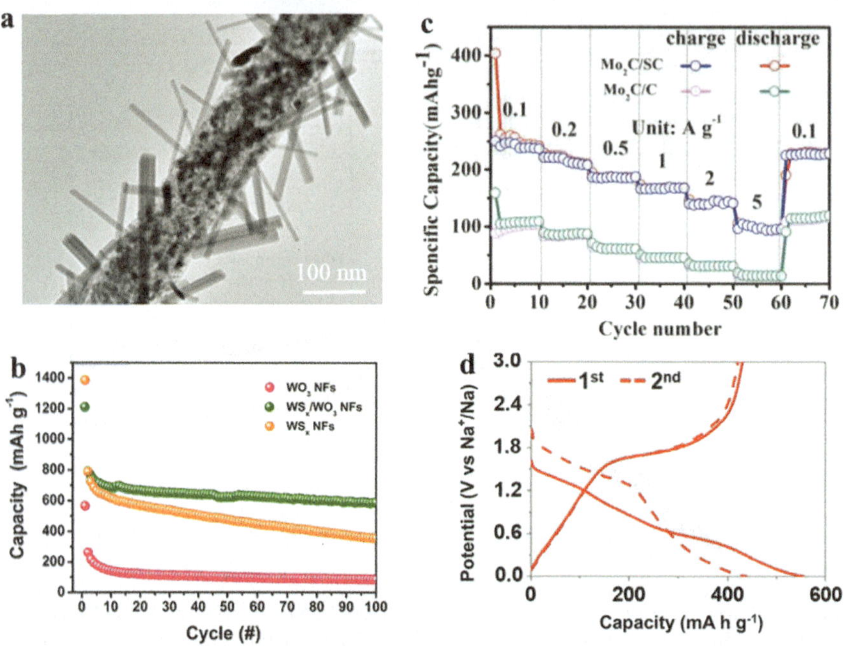

Fig. 3.11 **a** TEM image of WS$_x$ nanofibers at 400 °C in air. **b** Cycle performance of the WS$_x$ nanofibers and the post calcined WSx nanofibers at 400 and 500 °C at a current density of 100 mA g^{-1}. Reproduced from Ref. [68]. Copyright 2016, American Chemical Society. **c** Rate performance of the Mo$_2$C/SC and Mo$_2$C/C electrodes. Reproduced from Ref. [69]. Copyright 2018, Elsevier. **d** Discharge–charge curves of MC-NCNF/MoSe$_2$ at the current density of 0.5 A g^{-1}. Reproduced from Ref. [37]. Copyright 2019, Elsevier

composed of homogeneously dispersed single-crystalline NiSe$_2$ nanocrystals that have a mean size of 27 nm [70]. According to the Jung et al. statement, the design of ideal 2D transition metal dichalcogenide with CNF will improve the interlayer spacing resulting in high performance [67]. Similarly, Liu et al. reported Sb$_2$X$_3$ (X ¼ O, S)/carbon fiber improve the NIBs performance by attaining a stable capacity of 514 mAh g^{-1} after 500 cycles at 0.5A g^{-1} [71].

Also, Fig. 3.11c depicts the performance of Mo$_2$C/C with dopped nanofibers. Interestingly, Mo$_2$C/SC a superior rate performance and long-cycling stability over the 1000th cycles [69]. Very recently, Jeong et al. fabricated MC-NCNF/MoSe$_2$ nanofibers. This multi-channelled structure facilitated effective Na$^+$ and electron diffusion during repeated discharge/charge processes and accommodated the massive volume expansion. The two initial discharge–charge profiles shown in Fig. 3.11d were reliable with the CV curves. The initial discharge capacity and coulombic efficiency of MC-NCNF/MoSe$_2$ were 553 mAh g^{-1} and 77.3%, respectively. Additionally, the initial irreversible capacity loss was supplemented by SEI layer, which is derived from electrolyte decomposition and the carbonaceous material [37]. Table 3.1 summarizes the performance of various electrospun nanofibers and their hybrids

Table 3.1 The summary of various Electrospun nanofibers for NIBs application

Materials	Precursor	Process	NIB performance	References
G/C nanofibers	GO/PVP/H_2O	500 °C—1 h in Ar (90%:10%)	300.8 after 1000 cycles at 2000 mA g^{-1}	[42]
Hard carbon	PVC/THF, DMF (3.5: 6.5)	700 °C in Ar	211 mAh g^{-1} after 211 cycles at 12 mA g^{-1}	[72]
Fe_2O_3@C	$Fe(NO_3)_3 \cdot 9H_2O$, DMF/PAN	1st step: 260 °C—2 h in Ar (90%:10%) 2nd step: 290 °C—2 h in Ar/H_2 (90%:10%)	345 mAh g^{-1} after 1000 cycles at 2000 mA g^{-1}	[21]
Free-standing Porous CNFs	PAN, triblock copolymer Pluronic F127/DMF	1st step: 280 °C for 3 h in air	266 mAh g^{-1} after 100 cycles at 50 mA g^{-1}	[73]
Bi/C nanofibers	$BiCl_3$/$Bi(NO_3)_3$/DMF/PAN	1st step: 350 °C—2 h in Ar (90%:10%) 2nd step: 650 °C—3 h in Ar/H_2 (90%:10%)	273.2 mAh g^{-1} 500 cycles at 100 mA g^{-1}	[55]
SbNP@C	$SbCl_3$/PAN/DMF	1st step: 250 °C for 1 h in air 2nd step: 600 °C for 5 h in H_2/Ar (H_2, 5%)	350 mAh g^{-1} after 300 cycles At 100 mA g^{-1}	[52]
Sb/C/G	$SbCl_3$/PAN/DMF/GO	1st step: 280 °C for 2 h in Ar 2nd step: 600 °C for 3 h in H_2/Ar (1%:19%)	274 mAh g^{-1} after 100 cycles 100 mA g^{-1}	[74]
SnSb@C	SnO_2 NPs, Sb_2O_5 NPs/ PMMA, PAN/DMF	1st step: 280 °C for 5.5 h in air 2nd step: 700 °C for 3 h in air	356 mAh g^{-1} after 200 cycles At 500 mA g^{-1}	[75]
PCNF@SnO_2@C	PMMA, PAN/DMF	Electrospinning/electrodeposition	374 mAh g^{-1} after 100 cycles At 50 mA g^{-1}	[60]

(continued)

Table 3.1 (continued)

Materials	Precursor	Process	NIB performance	References
$Li_4Ti_5O_{12}$@C	$Li(ac)_2$.H_2O, $TiO(acac)_2$/ PVP/EtOH	1st step: 400 °C for 2 h in Ar 2nd step: 800 °C for 5 h in Ar	163 mAh g^{-1} after 100 cycles at 35 mA g^{-1}	[76]
$MnFe_2O_4$@C	$(Mn(CH_3COO)_2 \cdot 4H2O$, $Fe(NO_3)_3 \cdot 9H_2O$, /PAN/DMF	1st at 250 °C for 1 h in N_2 2nd 700 °C for 1 h in N_2	360 mAh g^{-1} after 4200 cycles at 2000 mA g^{-1}	[64]
MoS_2–C	$(NH_4)_2MoS_4$/PVP/DMF	1st step: 450 °C for 2 h H_2/Ar (H_2, 5%) 2nd step: 800 °C for 6 h in Ar	1007 mAh g^{-1} after 100 cycles at 1 mA g^{-1}	[77]
$NiSe_2$-rGO-C	$Ni(OCOCH_3)_2 \cdot$ $4H_2O$-PAN-PS-GO	1st 450 °C for 3 h under an Ar 300 °C for 10 h in H_2Se gas	468 after 100 cycles at 200 mA g^{-1}	[70]
TiO_2-coated MoS_2 NFs	$(NH_4)_2MoS_4$/SAN/DMF	900 °C for 6 h in H_2/N_2 (H_2, 5%)	474 mAh g^{-1} after 30 cycles at 100 mA g^{-1}	[78]
TiO_2@C	$Ti(OC_4H_9)_4$/PVP/ C_2H_5OH and CH_3COOH (10:1 by wt)	1st at 280 °C for 5.5 h in air 2nd 450, 550, 650 and 750 °C for 4 h	68.2% ICE after 100 cycles 30 mA g^{-1}	[66]

with metal/metal-oxides/conducting polymers/transition metal dichalcogenides and other carbon composites for high performance for NIBs. Moreover, we provided a detailed literature survey regarding the precursor salt, solvents and polymers used.

References

1. W. Luo, F. Shen, C. Bommier, H. Zhu, X. Ji, L. Hu, Na-ion battery anodes: materials and electrochemistry. Acc. Chem. Res. **49**(2), 231–240 (2016)
2. G.-L. Xu, R. Amine, A. Abouimrane, H. Che, M. Dahbi, Z.-F. Ma, I. Saadoune, J. Alami, W.L. Mattis, F. Pan, Z. Chen, K. Amine, Challenges in developing electrodes, electrolytes, and diagnostics tools to understand and advance sodium-ion batteries. Adv. Energy Mater. **8**(14) (2018)
3. N. Yabuuchi, K. Kubota, M. Dahbi, S. Komaba, Research development on sodium-ion batteries. Chem. Rev. **114**(23), 11636–11682 (2014)
4. M.D. Slater, D. Kim, E. Lee, C.S. Johnson, Sodium-ion batteries. Adv. Func. Mater. **23**(8), 947–958 (2013)
5. S. Li, Y. Dong, L. Xu, X. Xu, L. He, L. Mai, Effect of carbon matrix dimensions on the electrochemical properties of Na3V2 (PO4) 3 nanograins for high-performance symmetric sodium-ion batteries. Adv. Mater. **26**(21), 3545–3553 (2014)
6. M. Dahbi, N. Yabuuchi, K. Kubota, K. Tokiwa, S. Komaba, Negative electrodes for Na-ion batteries. Phys. Chem. Chem. Phys. **16**(29), 15007–15028 (2014)
7. C. Delmas, C. Fouassier, P. Hagenmuller, Structural classification and properties of the layered oxides. Phys. B+ c **99**(1–4), 81–85 (1980)
8. Y. Takeda, K. Nakahara, M. Nishijima, N. Imanishi, O. Yamamoto, M. Takano, R. Kanno, Sodium deintercalation from sodium iron oxide. Mater. Res. Bull. **29**(6), 659–666 (1994)
9. X. Xiang, K. Zhang, J. Chen, Recent advances and prospects of cathode materials for sodium-ion batteries. Adv. Mater. **27**(36), 5343–5364 (2015)
10. P. Poizot, F. Dolhem, Clean energy new deal for a sustainable world: from non-CO_2 generating energy sources to greener electrochemical storage devices. Energy Environ. Sci. **4**(6), 2003–2019 (2011)
11. S.W. Kim, D.H. Seo, X. Ma, G. Ceder, K. Kang, Electrode materials for rechargeable sodium-ion batteries: potential alternatives to current lithium-ion batteries. Adv. Energy Mater. **2**(7), 710–721 (2012)
12. Y. Chen, X. Hu, B. Evanko, X. Sun, X. Li, T. Hou, S. Cai, C. Zheng, W. Hu, G.D. Stucky, High-rate FeS2/CNT neural network nanostructure composite anodes for stable, high-capacity sodium-ion batteries. Nano Energy **46**, 117–127 (2018)
13. T. Jin, Y. Liu, Y. Li, K. Cao, X. Wang, L. Jiao, Electrospun NaVPO4F/C Nanofibers as self-standing cathode material for ultralong cycle life na-ion batteries. Adv. Energy Mater. **7**(15) (2017)
14. K. Kretschmer, B. Sun, J. Zhang, X. Xie, H. Liu, G. Wang, 3D interconnected carbon fiber network-enabled ultralong life Na3V2(PO4)(3)@Carbon paper cathode for sodium-ion batteries. Small **13**(9) (2017)
15. M. Mao, C. Cui, M. Wu, M. Zhang, T. Gao, X. Fan, J. Chen, T. Wang, J. Ma, C. Wang, Flexible ReS2 nanosheets/N-doped carbon nanofibers-based paper as a universal anode for alkali (Li, Na, K) ion battery. Nano Energy **45**, 346–352 (2018)
16. Y. Liu, N. Zhang, X. Liu, C. Chen, L.-Z. Fan, L. Jiao, Red phosphorus nanoparticles embedded in porous N-doped carbon nanofibers as high-performance anode for sodium-ion batteries. Energy Storage Mater. **9**, 170–178 (2017)
17. W. Luo, A. Calas, C. Tang, F. Li, L. Zhou, L. Mai, Ultralong Sb2Se3 nanowire-based free-standing membrane anode for lithium/sodium ion batteries. ACS Appl. Mater. Interfaces **8**(51), 35219–35226 (2016)

18. Y. Liu, N. Zhang, L. Jiao, J. Chen, Tin nanodots encapsulated in porous nitrogen-doped carbon nanofibers as a freestanding anode for advanced sodium-ion batteries. Adv. Mater. **27**(42), 6702+ (2015)
19. X. Guo, X. Zhang, H. Song, J. Zhou, Electrospun cross-linked carbon nanofiber films as free-standing and binder-free anodes with superior rate performance and long-term cycling stability for sodium ion storage. J. Mater. Chem. A **5**(40), 21343–21352 (2017)
20. L. Li, P. Liu, K. Zhu, J. Wang, G. Tai, J. Liu, Flexible and robust N-doped carbon nanofiber film encapsulating uniformly silica nanoparticles: Free-standing long-life and low-cost electrodes for Li-and Na-Ion batteries. Electrochim. Acta **235**, 79–87 (2017)
21. Y. Liu, F. Wang, L.-Z. Fan, Self-standing Na-storage anode of Fe_2O_3 nanodots encapsulated in porous N-doped carbon nanofibers with ultra-high cyclic stability. Nano Res. **11**(8), 4026–4037 (2018)
22. I. Kovalenko, B. Zdyrko, A. Magasinski, B. Hertzberg, Z. Milicev, R. Burtovyy, I. Luzinov, G. Yushin, A major constituent of brown algae for use in high-capacity Li-ion batteries. Science **334**(6052), 75–79 (2011)
23. B. Koo, H. Kim, Y. Cho, K.T. Lee, N.S. Choi, J. Cho, A highly cross-linked polymeric binder for high-performance silicon negative electrodes in lithium ion batteries. Angew. Chem. Int. Ed. **51**(35), 8762–8767 (2012)
24. T. Jin, Q. Han, L. Jiao, Binder-free electrodes for advanced sodium-ion batteries. Adv. Mater. 1806304 (2019)
25. H. Wang, X. Gao, J. Feng, S. Xiong, Nanostructured V2O5 arrays on metal substrate as binder free cathode materials for sodium-ion batteries. Electrochim. Acta **182**, 769–774 (2015)
26. Y. Sun, X. Liang, H. Xiang, Y. Yu, Free-standing vanadium pentoxide nanoribbon film as a high-performance cathode for rechargeable sodium batteries. Chin. Chem. Lett. **28**(12), 2251–2253 (2017)
27. S. Li, X. Li, Y. Li, B. Yan, X. Song, L. Fan, H. Shan, D. Li, Design of V2O5 · xH2O cathode for highly enhancing sodium storage. J. Alloy. Compd. **722**, 278–286 (2017)
28. E. Uchaker, Y. Zheng, S. Li, S. Candelaria, S. Hu, G. Cao, Better than crystalline: amorphous vanadium oxide for sodium-ion batteries. J. Mater. Chem. A **2**(43), 18208–18214 (2014)
29. E. Brown, J. Acharya, A. Elangovan, G.P. Pandey, J. Wu, J. Li, Disordered bilayered V2O5 · nH2O shells deposited on vertically aligned carbon nanofiber arrays as stable high-capacity sodium ion battery cathodes·. Energy Technol. **6**(12), 2438–2449 (2018)
30. S. Liu, Z. Tong, J. Zhao, X. Liu, X. Ma, C. Chi, Y. Yang, X. Liu, Y. Li, Rational selection of amorphous or crystalline V 2 O 5 cathode for sodium-ion batteries. Phys. Chem. Chem. Phys. **18**(36), 25645–25654 (2016)
31. S. Petnikota, R. Chua, Y. Zhou, E. Edison, M. Srinivasan, Amorphous vanadium oxide thin films as stable performing cathodes of lithium and sodium-ion batteries. Nanoscale Res. Lett. **13**(1), 363 (2018)
32. C.G. Hawkins, L. Whittaker-Brooks, Vertically oriented TiS2−x nanobelt arrays as binder- and carbon-free intercalation electrodes for Li-and Na-based energy storage devices. J. Mater. Chem. A **6**(44), 21949–21960 (2018)
33. H.J. Yoon, M.E. Lee, N.R. Kim, S.J. Yang, H.-J. Jin, Y.S. Yun, Hierarchically nanoporous pyropolymer nanofibers for surface-induced sodium-ion storage. Electrochim. Acta **242**, 38–46 (2017)
34. Y. Li, C. Liu, Z. Xie, J. Yao, G. Cao, Superior sodium storage performance of additive-free V_2O_5 thin film electrodes. J. Mater. Chem. A **5**(32), 16590–16594 (2017)
35. J. Liu, J. Dai, L. Huang, B. Fu, Flexible and binder-free electrospun Co_3O_4 nanoparticles/carbon composite nanofiber mats as negative electrodes for sodium-ion batteries. Funct. Mater. Lett. **11**(4) (2018)
36. S. Yuan, X.l. Huang, D.l. Ma, H.G. Wang, F.Z. Meng, X.B. Zhang, Engraving copper foil to give large-scale binder-free porous CuO arrays for a high-performance sodium-ion battery anode. Adv. Mater. **26**(14), 2273–2279 (2014)
37. S.Y. Jeong, S. Ghosh, J.-K. Kim, D.-W. Kang, S.M. Jeong, Y.C. Kang, J.S. Cho, Multi-channel-contained few-layered MoSe2 nanosheet/N-doped carbon hybrid nanofibers prepared using

diethylenetriamine as anodes for high-performance sodium-ion batteries. J. Ind. Eng. Chem. **75**, 100–107 (2019)

38. H. Yin, M.-L. Cao, X.-X. Yu, H. Zhao, Y. Shen, C. Li, M.-Q. Zhu, Self-standing Bi_2O_3 nanoparticles/carbon nanofiber hybrid films as a binder-free anode for flexible sodium-ion batteries. Mater. Chem. Front. **1**(8), 1615–1621 (2017)

39. X. Xia, J. Xie, S. Zhang, B. Pan, G. Cao, X. Zhao, Wrinkled graphene-reinforced nickel sulfide thin film as high-performance binder-free anode for sodium-ion battery. J. Mater. Sci. Technol. **33**(8), 775–780 (2017)

40. X. Xia, J. Xie, S. Zhang, B. Pan, G. Cao, X. Zhao, Ni_3S_2 nanosheet-anchored carbon submicron tube arrays as high-performance binder-free anodes for Na-ion batteries. Inorganic Chem. Front. **4**(1), 131–138 (2017)

41. H. Yin, H.-Q. Qu, Z. Liu, R.-Z. Jiang, C. Li, M.-Q. Zhu, Long cycle life and high rate capability of three dimensional CoSe2 grain-attached carbon nanofibers for flexible sodium-ion batteries. Nano Energy **58**, 715–723 (2019)

42. Y. Liu, L.-Z. Fan, L. Jiao, Graphene highly scattered in porous carbon nanofibers: a binder-free and high-performance anode for sodium-ion batteries. J. Mater. Chem. A **5**(4), 1698–1705 (2017)

43. W. Zheng, X.B. Huang, Y.R. Ren, H.Y. Wang, S.B.A. Zhou, Y.D. Chen, X. Ding, T. Zhou, Porous spherical Na3V2 (PO4)(3)/C composites synthesized via a spray drying-assisted process with high-rate performance as cathode materials for sodium-ion batteries. Solid State Ion. **308**, 161–166 (2017)

44. Y. Hu, L. Wu, G. Liao, Y. Yang, F. Ye, J. Chen, X. Zhu, S. Zhong, Electrospinning synthesis of Na2MnPO4F/C nanofibers as a high voltage cathode material for Na-ion batteries. Ceram. Int. **44**(15), 17577–17584 (2018)

45. X. Miao, Z. Chen, N. Wang, Y. Nuli, J. Wang, J. Yang, S.-I. Hirano, Electrospun V2MoO8 as a cathode material for rechargeable batteries with Mg metal anode. Nano Energy **34**, 26–35 (2017)

46. Y. Niu, M. Xu, C. Dai, B. Shen, C.M. Li, Electrospun graphene-wrapped Na6.24Fe4.88(P2O7)(4) nanofibers as a high-performance cathode for sodium-ion batteries. Phys. Chem. Chem. Phys. **19**(26), 17270–17277 (2017)

47. J.-C. Kim, D.-W. Kim, Electrospun Cu/Sn/C nanocomposite fiber anodes with superior usable lifetime for lithium- and sodium-ion batteries. Chem. Asian J. **9**(11), 3313–3318 (2014)

48. J.-C. Kim, D.-W. Kim, Synthesis of multiphase SnSb nanoparticles-on-SnO2/Sn/C nanofibers for use in Li and Na ion battery electrodes. Electrochem. Commun. **46**, 124–127 (2014)

49. L. Ji, M. Gu, Y. Shao, X. Li, M.H. Engelhard, B.W. Arey, W. Wang, Z. Nie, J. Xiao, C. Wang, J.-G. Zhang, J. Liu, Controlling SEI formation on SnSb-Porous carbon nanofibers for improved Na ion storage. Adv. Mater. **26**(18), 2901–2908 (2014)

50. H. Jia, M. Dirican, C. Chen, J. Zhu, P. Zhu, C. Yan, Y. Li, X. Dong, J. Guo, X. Zhang, Reduced graphene oxide-incorporated SnSb@CNF composites as anodes for high-performance sodium-ion batteries. ACS Appl. Mater. Interfaces **10**(11), 9696–9703 (2018)

51. L. Wu, X. Hu, J. Qian, F. Pei, F. Wu, R. Mao, X. Ai, H. Yang, Y. Cao, Sb–C nanofibers with long cycle life as an anode material for high-performance sodium-ion batteries. Energy Environ. Sci. **7**(1), 323–328 (2014)

52. Y. Zhu, X. Han, Y. Xu, Y. Liu, S. Zheng, K. Xu, L. Hu, C. Wang, Electrospun Sb/C fibers for a stable and fast sodium-ion battery anode. ACS Nano **7**(7), 6378–6386 (2013)

53. M. Zhu, X. Kong, H. Yang, T. Zhu, S. Liang, A. Pan, One-dimensional coaxial Sb and carbon fibers with enhanced electrochemical performance for sodium-ion batteries. Appl. Surf. Sci. **428**, 448–454 (2018)

54. M. Sha, H. Zhang, Y. Nie, K. Nie, X. Lv, N. Sun, X. Xie, Y. Ma, X. Sun, Sn nanoparticles@ nitrogen-doped carbon nanofiber composites as high-performance anodes for sodium-ion batteries. J. Mater. Chem. A **5**(13), 6277–6283 (2017)

55. H. Yin, Q. Li, M. Cao, W. Zhang, H. Zhao, C. Li, K. Huo, M. Zhu, Nanosized-bismuth-embedded 1D carbon nanofibers as high-performance anodes for lithium-ion and sodium-ion batteries. Nano Res. **10**(6), 2156–2167 (2017)

56. J. Liang, X. Gao, J. Guo, C. Chen, K. Fan, J. Ma, Electrospun MoO2@ NC nanofibers with excellent Li+/Na+ storage for dual applications. Sci. China Mater. **61**(1), 30–38 (2018)
57. Y. Wu, Y. Jiang, J. Shi, L. Gu, Y. Yu, Multichannel porous TiO2 hollow nanofibers with rich oxygen vacancies and high grain boundary density enabling superior sodium storage performance. Small **13**(22), 1700129 (2017)
58. B. Zhang, J. Huang, J.K. Kim, Ultrafine amorphous SnOx embedded in carbon nanofiber/carbon nanotube composites for li-ion and na-ion batteries. Adv. Func. Mater. **25**(32), 5222–5228 (2015)
59. Z.-L. Xu, S. Yao, J. Cui, L. Zhou, J.-K. Kim, Atomic scale, amorphous FeOx/carbon nanofiber anodes for Li-ion and Na-ion batteries. Energy Storage Mater. **8**, 10–19 (2017)
60. M. Dirican, Y. Lu, Y. Ge, O. Yildiz, X. Zhang, Carbon-Confined Sno(2)-electrodeposited porous carbon nanofiber composite as high-capacity sodium-ion battery anode material. ACS Appl. Mater. Interfaces **7**(33), 18387–18396 (2015)
61. Y. Xiong, J. Qian, Y. Cao, X. Ai, H. Yang, Electrospun TiO2/C nanofibers as a high-capacity and cycle-stable anode for sodium-ion batteries. ACS Appl. Mater. Interfaces **8**(26), 16684–16689 (2016)
62. L. Li, Z. Chen, H. Zhang, Z. Zhu, M. Zhang, The double effects of sulfur-doping on MoO2/C nanofibers with high properties for Na-ion batteries. Appl. Surf. Sci. **455**, 343–348 (2018)
63. S. Nie, L. Liu, J. Liu, J. Xia, Y. Zhang, J. Xie, M. Li, X. Wang, TiO2-Sn/C composite nanofibers with high-capacity and long-cycle life as anode materials for sodium ion batteries. J. Alloy. Compd. **772**, 314–323 (2019)
64. Y. Liu, N. Zhang, C. Yu, L. Jiao, J. Chen, MnFe2O4@ C nanofibers as high-performance anode for sodium-ion batteries. Nano Lett. **16**(5), 3321–3328 (2016)
65. L. Wu, J. Lang, R. Wang, R. Guo, X. Yan, Electrospinning synthesis of mesoporous MnCoNiO x@ double-carbon nanofibers for sodium-ion battery anodes with pseudocapacitive behavior and long cycle life. ACS Appl. Mater. Interfaces **8**(50), 34342–34352 (2016)
66. Y. Ge, J. Zhu, Y. Lu, C. Chen, Y. Qiu, X. Zhang, The study on structure and electrochemical sodiation of one-dimensional nanocrystalline TiO2@ C nanofiber composites. Electrochim. Acta **176**, 989–996 (2015)
67. J.-W. Jung, W.-H. Ryu, S. Yu, C. Kim, S.-H. Cho, I.-D. Kim, Dimensional effects of MoS2 nanoplates embedded in carbon nanofibers for bifunctional Li and Na insertion and conversion reactions. ACS Appl. Mater. Interfaces **8**(40), 26758–26768 (2016)
68. W.-H. Ryu, H. Wilson, S. Sohn, J. Li, X. Tong, E. Shaulsky, J. Schroers, M. Elimelech, A.D. Taylor, Heterogeneous WS x/WO3 thorn-bush nanofiber electrodes for sodium-ion batteries. ACS Nano **10**(3), 3257–3266 (2016)
69. L. Li, Z. Chen, M. Zhang, Mo2C embedded in S-doped carbon nanofibers for high-rate performance and long-life time Na-ion batteries. Solid State Ion. **323**, 151–156 (2018)
70. J.S. Cho, S.Y. Lee, Y.C. Kang, First introduction of NiSe 2 to anode material for sodium-ion batteries: a hybrid of graphene-wrapped NiSe 2/C porous nanofiber. Sci. Rep. **6**, 23338 (2016)
71. S.A. Liu, Z.Y. Cai, J. Zhou, M.N. Zhu, A.Q. Pan, S.Q. Liang, High-performance sodium-ion batteries and flexible sodium-ion capacitors based on Sb2X3 (X = O, S)/carbon fiber cloth. J. Mater. Chem. A **5**(19), 9169–9176 (2017)
72. Y. Bai, Z. Wang, C. Wu, R. Xu, F. Wu, Y. Liu, H. Li, Y. Li, J. Lu, K. Amine, Hard carbon originated from polyvinyl chloride nanofibers as high-performance anode material for Na-ion battery. ACS Appl. Mater. Interfaces **7**(9), 5598–5604 (2015)
73. W. Li, L. Zeng, Z. Yang, L. Gu, J. Wang, X. Liu, J. Cheng, Y. Yu, Free-standing and binder-free sodium-ion electrodes with ultralong cycle life and high rate performance based on porous carbon nanofibers. Nanoscale **6**(2), 693–698 (2014)
74. K. Li, D. Su, H. Liu, G. Wang, Antimony-carbon-graphene fibrous composite as freestanding anode materials for sodium-ion batteries. Electrochim. Acta **177**, 304–309 (2015)
75. C. Chen, K. Fu, Y. Lu, J. Zhu, L. Xue, Y. Hu, X. Zhang, Use of a tin antimony alloy-filled porous carbon nanofiber composite as an anode in sodium-ion batteries. RSC Adv. **5**(39), 30793–30800 (2015)

76. J. Liu, K. Tang, K. Song, P.A. van Aken, Y. Yu, J. Maier, Tiny Li 4 Ti 5 O 12 nanoparticles embedded in carbon nanofibers as high-capacity and long-life anode materials for both Li-ion and Na-ion batteries. Phys. Chem. Chem. Phys. **15**(48), 20813–20818 (2013)
77. C. Zhu, X. Mu, P.A. van Aken, Y. Yu, J. Maier, Single-layered ultrasmall nanoplates of MoS2 embedded in carbon nanofibers with excellent electrochemical performance for lithium and sodium storage. Angew. Chem. Int. Ed. **53**(8), 2152–2156 (2014)
78. W.-H. Ryu, J.-W. Jung, K. Park, S.-J. Kim, I.-D. Kim, Vine-like MoS2 anode materials self-assembled from 1-D nanofibers for high capacity sodium rechargeable batteries. Nanoscale **6**(19), 10975–10981 (2014)

Chapter 4
Electrospinning of Nanofibers for K-Ion Battery

Abstract Alkali ion batteries have virtually been commonly used as a power supply for portable electronic items and electric vehicles. Yet, unavailability of conventional anodes for Li, Na and K ion battery are impeding their development, notably for Na-ion battery and K-ion battery. Potassium-ion batteries have been viewed as the most encouraging replacement for LIBs owing to the benefits of cheap, nontoxicity, large quantity, and low redox potential of K/K^+ [2.93 V versus standard hydrogen electrode (SHE)]. However, the main disadvantages of larger ionic diameter hinder the insertion of K-ions into the electrode. In this chapter, we include the critical technological developments and systematic experiments for a comprehensive range of rechargeable potassium batteries. Mainly, we highlighted the electrospun nanofibers-based electrodes with valuable research which can enhance the rechargeable K-ion batteries.

Keywords K-ion battery · Fiber electrode · Flexibility · Electrochemical properties

4.1 K-Ion Battery Working Principle and Cell Structure

K-ion battery is shared the same "rocking chair" principle of set-up as their LIBs analogues. K-ion batteries are a notable alternative to Li and Na-ion batteries since K-element is rich in the earth like Na-element. K-ion battery composed of four main parts of anode, cathode, separator and electrolyte as demonstrated in Fig. 4.1. In a model K-ion battery, graphite used as an anode, K-MnHCFe was acting as a cathode, glass fiber filter used as a separator and the 0.7 M KPF_6 EC/DEC (1:1v/v) used as an electrolyte.

In general, the standard potential of K^+/K (2.88 V vs. the SHE) is even 0.09 V smaller than that of Li^+/Li (2.79 V vs. the SHE) in propylene carbonate (PC). Therefore, it is powerful for high voltage operation. Besides, K-ions mobility in the electrolyte and electrode/electrolyte interface is expected to be higher due to the weaker Lewis acidity of K^+ ions than other metal ions such as Li^+, Mg^{2+}, and Na^+ [1]. This K-ion battery received much attention, but they are less studied and are in the earlier stage of progress.

S. Peng and P. R. Ilango, *Electrospinning of Nanofibers for Battery Applications*,
https://doi.org/10.1007/978-981-15-1428-9_4

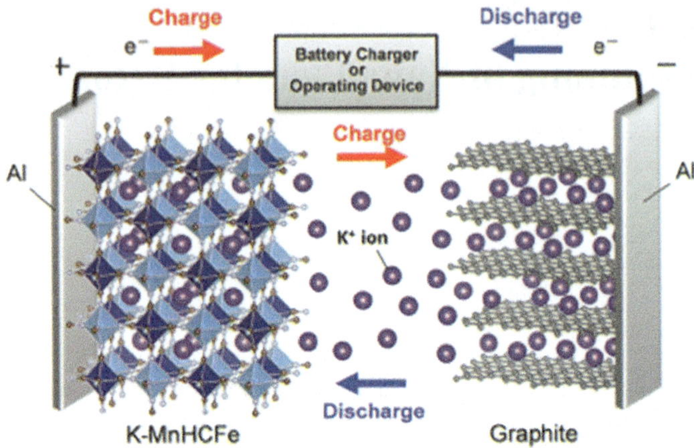

Fig. 4.1 Schematic representation of K-ion full cell battery. Reproduced from Ref. [1]. Copyright 2017, The Royal Society of Chemistry

4.2 Perspectives on Material Development

4.2.1 Cathode

Finding new kind of materials includes cathode materials since cathodes also contribute to the energy density of the battery and a wide range of research activities have been addressed in recent years in the development of cathode materials. Concerning cathode materials for K-ion batteries, layered $K_{0.3}MnO_2$, poly-anionic $KFeSO_4F$, amorphous $FePO_4$, and organic perylene-tetracarboxylic acid-dianhydride have been reported [2–5]. Alternatively, Prussian blue analogues considered suitable cathode materials to diffuse ions with large size. Prussian blue analogues denoted as $A_xM1[M2(CN)_6]_ynH_2O$, where A is a mobile metal and M1 and M2 are transition metals. Assuming, both transition metals participate electrochemical redox of $M1^{2+/3+}$ and $M2^{2+/3+}$ couples, the theoretical capacity is 155 mAh g^{-1} by supposing that x in $K_xFe[Fe(CN)_6]$ varies from 2 to 0 [1].

4.2.2 Anode

At present, the new thought of K-ion battery has great interest due to the cheap K resources. However, the higher K-ionic radius (0.138 nm) making trouble to intercalate into the crystal structure. To resolve this crisis, material development for both anode and cathodes is essential but limited at current achievement. Pioneer work graphite had been used as an anode material with obvious voltage plateau and a good theoretical capacity of ~278 mAh g^{-1} [6, 7]. Nevertheless, graphite was not suitable anode since its failure in NIBs; Therefore, K-intercalated graphite was performed in molten KF or KF/AlF$_3$ at 750 °C [8]. Later, K-inserted non-graphitic carbon nanofibers were used at room RT [9]. KC$_8$, the stage-one K-GIC, as the first figured out alkali metal GIC, has been easily prepared by potassium vapour transport or by soaking graphite in a non-aqueous solution with potassium metal solvated [9]. Recently, other anode materials (Si, Sn, Sb, Bi and P) also proposed which can alloy with K and reveals good capacity and stability [10, 11]. The theoretical study says that Sn and Si are appropriate anode materials for KIBs and they can alloy with K to form KSn and KSi to bring capacities of 226 and 955 mAh g^{-1} at an average voltage of 0.32 and 0.15 V corresponding to K/K$^+$, respectively [12]. On the other hand, transition metal dichalcogenides offer a wide range of interlayer spacing to intercalate K-ions. For example, MoS$_2$ is plentiful, which claim alternative anode for future anode materials due to the formation of K$_{0.4}$ MoS$_2$ after insertion of K$^+$ which does not affect the original hexagonal structure of MoS$_2$ [13]. Likewise, MoSe$_2$ has been involved alternative the anode material for KIBs since its larger interlayer spacing allows more K-ions during the electrochemical reaction which leads to excellent rate capability, and long term cycling stability [14]. Similarly, other TMDs like CoS [15] SnS$_2$ [16] and Sb$_2$S$_3$ [17] exhibits enhanced K-ion storage performance. Alternatively, carbon and carbon-based composites such as hard carbon [18] soft carbon [19], graphene [20], Biocarbon [21] have been utilized as one of the most encouraging electrode materials because of reasonable K storage capability and good thermal stability. Besides, specific energy density is another problem to be the commercialization of K-ion battery. The electrochemical performance of a different of anode materials for K-ion batteries are condensed in Table 4.1

4.3 Free-Standing Electrodes

4.3.1 Free-Standing Anodes

Zhao et al. synthesized free-standing nitrogen-doped cup-stacked carbon nanotube (NCSCNT) mats with nanosized feature and interconnected flexible structure, which possesses a reversible capacity of 323 mAh g^{-1} and notably improved rate capability retaining 75 mAh g^{-1} even at 1000 mA/g [32]. Very recent past, Yang et al. have drawn much attention to developing free-standing film electrodes. The prepared necklace-like hollow carbon film displays micro/meso/macro-pores, ultra-high pyrrolic/pyridinic-N doping and a high specific surface area, which could finally promote the intercalation/deintercalation of K$^+$, reduce the volume expansion and improve the stability of KIBs [33]. These findings reveal a high reversible specific capacity of 293.5 mAh g^{-1} at 100 mA g^{-1}, outstanding rate performance (204.8 mAh g^{-1} at 2000 mA g^{-1}) and cycling performance (161.3 mA h g^{-1} at 1000 mA g^{-1} after 1600 cycles), which indicate the excellent performance for carbon-based non-metal materials for KIBs. V$_2$O$_3$ is considered as substantial electrode material due

Table 4.1 Different of anode materials for K-ion batteries

Anode	Mechanism	Cutoff	Initial capacity	1st CE (%)	References
Graphite	Insertion	0.01–1.5 °C	273 mAh/g	57.4	[9]
Activated carbon	Insertion	0.01–2.0 V	260 mAh/g	62	[22]
Hard carbon	Insertion	0.1–1.0 V	200 mAh/g	50	[2]
RGO	Insertion	0.01–2.0 V	207 mAh/g	73.9	[7]
Porous carbon	Insertion	0–3.0 V	272 mAh/g	24.1	[23]
N, O doped CNF	Insertion	0.005–3 V	280 mAh/g	31	[24]
$K_2Ti_4O_9$	Insertion	0.01–2.5 V	97 mAh/g	20	[25]
Ti_3C_2	Insertion	0.01–3.5 V	146 mAh/g	43	[26]
MoS_2	Insertion	0.5–2.0 V	98 mAh/g	74.4	[27]
Co_3O_4–Fe_2O_3/C	Conversion	0.01–3.0 V	420 mAh/g	54	[28]
SnS_2-RGO	Conversion/alloying	0.01–2.0 V	355 mAh/g	56	[29]
Sb@NP/C	Alloying	0.001–2.5 V	596.8	46.2	[30]
Sn/C	Alloying	0.01–2.0 V	150	51.4	[31]

to its intrinsic tunnel structure provided by 3D V-V framework. Jin et al. approached V_2O_3@PNCNFs as a flexible and self-standing electrode in which of V_2O_3 nanoparticles embedded in porous N-doped CNF. Figure 4.2a illustrates the SEM images of V_2O_3@PNCNFs electrode. The DFT calculation suggests that the potassium storage of V_2O_3 is dominated by intercalation pseudocapacitance and only up to 1 mol K^+ can insert into V_2O_3 crystal. The first time reported and extensively cleared the mechanism during the intercalation that K^+ intercalate into the tunnels of V_2O_3 accompanied by a faradaic charge transfer with no crystallographic phase change. Subsequent electrochemical studies disclose three important mechanisms; (i) Oxide-based self- standing electrode maintains good rate capability shown in Fig. 4.2b. (ii) The capacitive charge-storage occupies a relatively large portion of the total capacity, and they confirmed that the pseudocapacitive process is the main reaction for energy storage, while the diffusion-controlled K^+ insertion is in the minority. (iii) It can create extrinsic pseudocapacitance by increasing potassium ions storage sites on the surface. Therefore, the pseudocapacitive process is beneficial to high-rate potassium storage [34]. Very recently, Wu et al. proposed a red P@N-PHCNFs free-standing electrode in which red Phosphorous embedded into nitrogen-doped porous hollow carbon nanofibers. The formation of P–C chemical bonds and the N-doping with

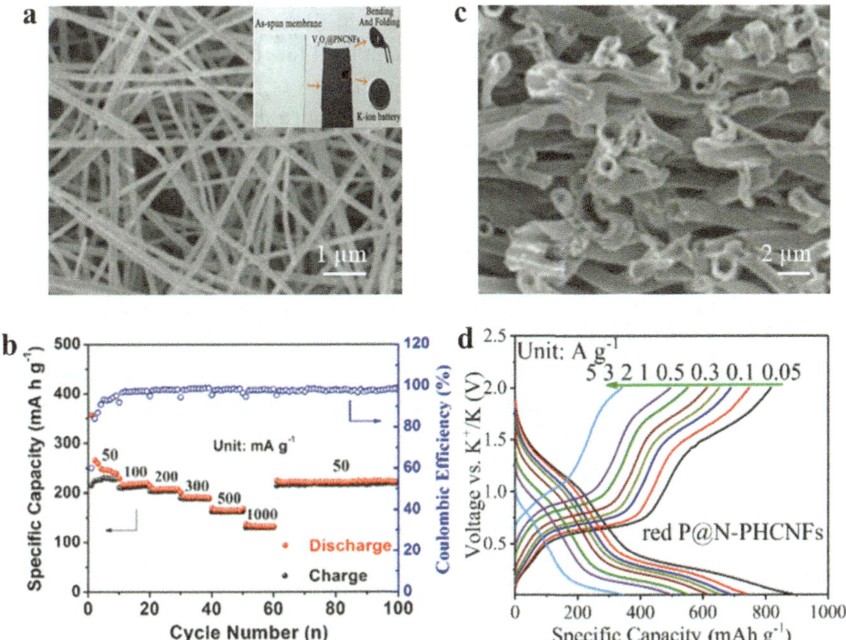

Fig. 4.2 **a** SEM and Digital photos of V_2O_3@PNCNFs electrode. **b** Rate performance of V_2O_3@PNCNFs electrode. Reproduced from Ref. [34]. Copyright 2018, Elsevier. **c** SEM cross-section images red P@N-PHCNFs sample. **d** Corresponding charge/discharge profiles of the red P@N-PHCNFs electrode at various current densities. Reproduced from Ref. [35]. Copyright 2019, American Chemical Society

boosted adsorption energy for red P could commendably relieve the huge volume change. Raman and ex situ XRD results, the final potassiation product of red P is anticipated to be K_4P_3. Thus, the reaction mechanism of red P is $4K + 3P \leftrightarrows K_4P_3$, agreeing to 1154 mAh g⁻¹ theoretical capacity of red P. The SEM cross-section image displayed in Fig. 4.2c which implies no noticeable red P particles could be detected on the surface, signifying that red P is well encapsulated into the 3D interconnected NPHCNFs matrix. Moreover, the charge–discharge profiles maintain the identical shape, with slight shifting and small overpotential, designating the high reversibility at high current densities [35] shown in Fig. 4.2d.

4.4 Binder-Free Electrodes

One dimensional carbon nanofibers (CNF) possesses excellent outcomes for enhancing the capacity and stability of Li, Na and K-ion batteries due to their shortened pathway for ion transport, good conductivity, and superb stress tolerance. To the extent, further improving K-ion electrochemical performance, its ultimate to design binder-free electrodes with nanostructures, in which all the materials could contribute to charge storage. This construction not only can facilitate cell packing by removing inactive additives such as binders and current collectors but also assure upgraded overall performance when the total volume of the device is taken into account.

4.4.1   Binder-Free Cathodes

Even though binder-free electrodes have undergone extensive expansion for LIBs [36] and NIBs [37], recently few research has been done for KIBs. For example, Zhu et al. developed a facile route to fabricate flexible binder-free KIB electrodes from Rusty Stainless-steel meshes (RSSM). When combined with RGO, the as-prepared RGO@PB@SSM electrodes delivers outstanding electrochemical behaviour for KIBs, including the high capacity of 96.8 mAh g^{-1}, high discharge voltage 3.3 V, high rate capability (1000 mAg^{-1}; 42% capacity retention) [38]. Figure 4.3a, b shows the potassium transport processes and cycle life of RGO@PB@SSM binder-free cathodes. The nanostructured binder-free anode has also developed for potential KIBs due to the material flexibility and structural integrity.

4.4.2   Binder-Free Anodes

Adams et al. proposed carbon nanofibers fabricated by electrospinning of polyacrylonitrile polymer and subsequent carbonization process. The influence of oxygen functionalization on K-ion carbon anode performance was tested for the first time via plasma oxidation of prepared carbon nanofibers. The binder-free mat (CNF-O) material consisted of a smooth morphology of randomly distributed fibers with an average diameter of 290 ± 39 nm shown in Fig. 4.4a. The energetic plasma oxidation process partially etched away from the surface carbon layer through ion bombardment, removing surface atoms and increasing surface roughness. As prepared fibers hold outstanding cycling stability through the amorphous carbon structuring and one-dimensional architecture, resulting in a stable capacity of 170 mAh g^{-1} after 1900 cycles at 1 °C rate for N-rich carbon nanofibers shown in Fig. 4.4b. Furthur the diffusion coefficient was calculated by Ficks second law

$$D = \frac{4}{\pi\tau}\left(\frac{m_B}{\rho S}\right)^2\left(\frac{\Delta E_s}{\Delta E_t}\right)^2 \tag{4.1}$$

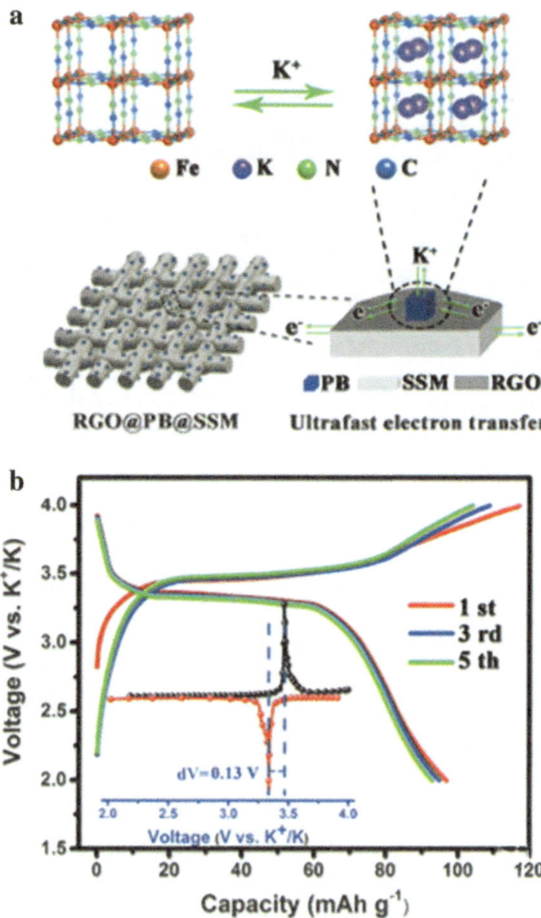

Fig. 4.3 **a** Potassium transport processes for RGO@PB@SSM electrodes. **b** Galvanostatic charge and discharge curves of RGO@PB@SSM. Reproduced from Ref. [38], Copyright 2017 Wiley–VCH

where τ is the pulse length, m_B is the electrode active mass, ρ is the material density, S is the surface area of the electrode calculated from BET results, ΔE_s is the steady-state voltage change due to the pulse, and ΔE_t is the voltage change during the constant current pulse [24]. Later, Mao et al. reported Rhenium disulfide (ReS$_2$), as binder and collector free-anode material. As fabricated ReS$_2$/N-CNFs paper consists of uniform, straight nanofibers with a diameter of 200–400 nm provided in Fig. 4.4c. The result concludes that ReS$_2$/N-CNFs provides of large interlayer spacing and extremely weakly van der Waals interaction of ReS$_2$, high conductivity and oriented electronic/ionic transport pathway of CNFs, and strong absorption of N-doping to sulfur and polysulfide, thus showing superb electrochemical characteristics in KIBs. It delivers reversible capacities of 253 mA h/g after 100 cycles at 50 mA/g in KIBs [39] see in Fig. 4.4d.

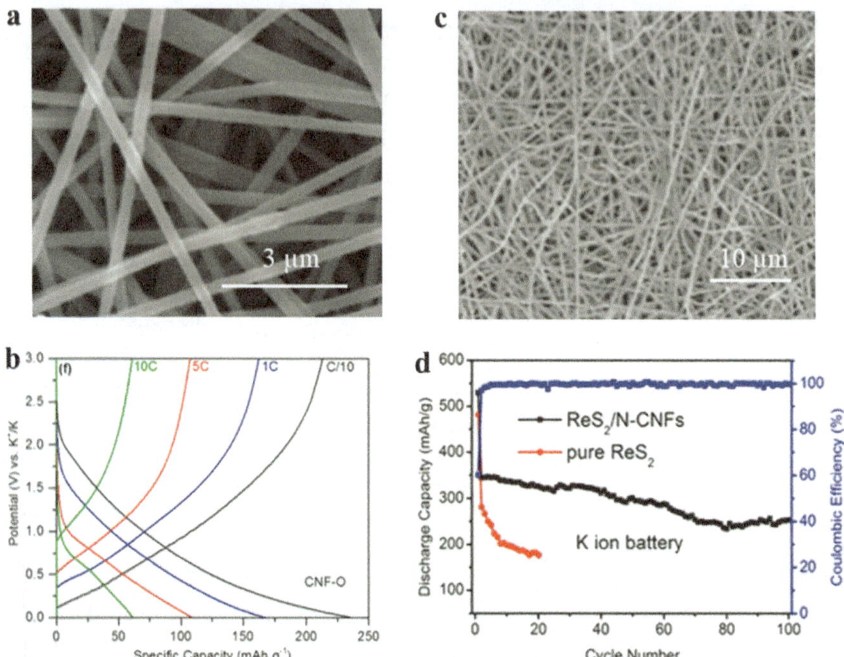

Fig. 4.4   **a** SEM image of CNF post oxygen plasma treatment(CNF-O). **b** Charge–discharge curves of CNF-O. **c** SEM of ReS$_2$/N-CNFs. **d** Cycling performance of the electrodes based on ReS$_2$/N-CNFs and pure ReS$_2$ at a current density of 100 mA/g. Reproduced from Ref. [39]. Copyright 2018, Elsevier

4.5   Other Electrodes for Anode

4.5.1   Metal/CNF Composites

During the past decades, the carbonaceous materials have considered potential anode candidate for alkali-metal ion batteries, Ex., LIBs, NIB and KIBs. Owing to the cheap and abundance in nature makes its successful implementation for potential applications. They demonstrate significant capacity but display poor rate capability and fast capacity decay. Fibrous carbon materials have drawn the observation of scientists and engineers. During the last decades, electrospun nanometer-to sub-micrometre-sized polymer nanofibers have captivated much attention in both research and commerce [40]. Liu et al. prepared carbon nanofibers, and they have discussed the electrochemical behaviour of Na and K-ion insertion via in-situ TEM analysis. They confirm the electrochemical reactions and corresponding structure changes of hollow bilayer CNFs during sodiation and potassiation. The remarkable point discussed is that d-C exhibits about three times the volume expansion of c–C, thus indicating a larger Na or K ion storage capability in d-C than c–C. Moreover, the longitudinal cracks are often seen near the c–C/d-C interface after sodiation and potassiation. Surprisingly, shrinkage of the inner tube is explained in terms of anisotropic sodiation strains in c–C and d-C [41]. The TEM images and corresponding charge–discharge and cycling behaviour for KIBs illustrated in Fig. 4.5a–d.

Besides, N-doped carbon materials have been studied to enhance K-ion storage performance because of the optimized electronic structure by N-doping. The enlarging the interlayer spacing helps the faster of K^+ and tolerates the volume expansion during the electrochemical reaction. Not long ago, Lv et al. attempted S/N co-doped carbon nanofiber aerogels through a pyrolysis sustainable seaweed (Fe-alginate) aerogel approach. The S/N-CNFAs electrode delivers high reversible capacities of 356 and 112 mAh g^{-1} at 100 and 5000 mA g^{-1}, respectively. Moreover, the DFT investigations pointed out that S/N co-doping is helpful to enhance K ions storage and reducing the diffusion barrier of K ions [42]. Meanwhile, Zhang et al. attempted to produce N-doped mesoporous carbon nanofibers (NMCNFs) by using facile electrospinning then followed with "high-temperature self-fluid technology". Schematic illustration of the fabrication procedure of NMCNFs shown Fig. 4.6a.

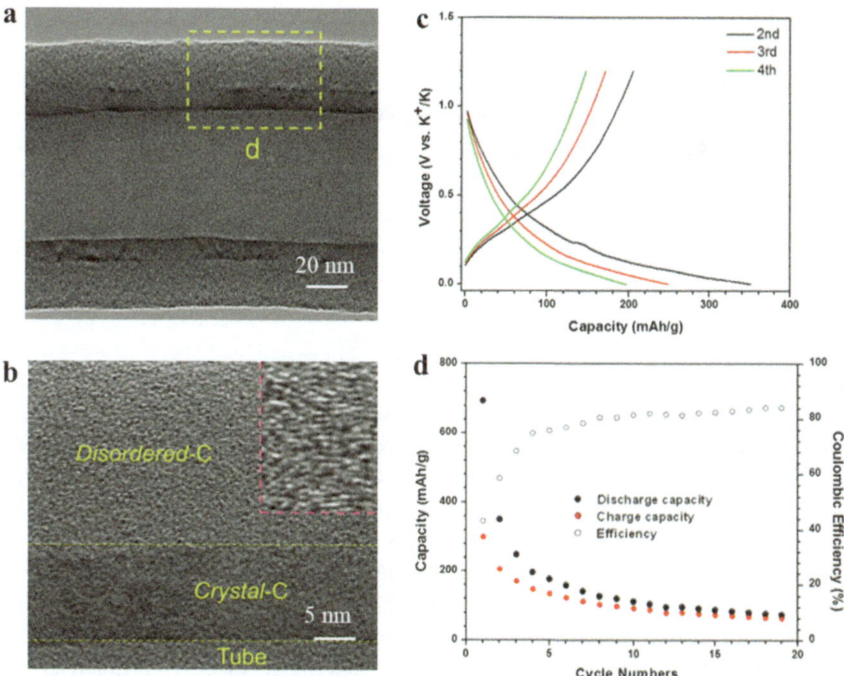

Fig. 4.5 **a** Typical morphology and structure of a pristine hollow CNF with a bilayer wall. **b** HRTEM image of the CNF with the inset showing the local graphite lattice fringes in the d-C layer. **c** Galvanostatic charge–discharge profiles in the second, third, and fourth cycle in KIBs. **d** Cycling performance and Coulombic efficiency of bilayer CNF electrodes in KIBs. Reproduced from Ref. [41]. Copyright 2014, American Chemical Society

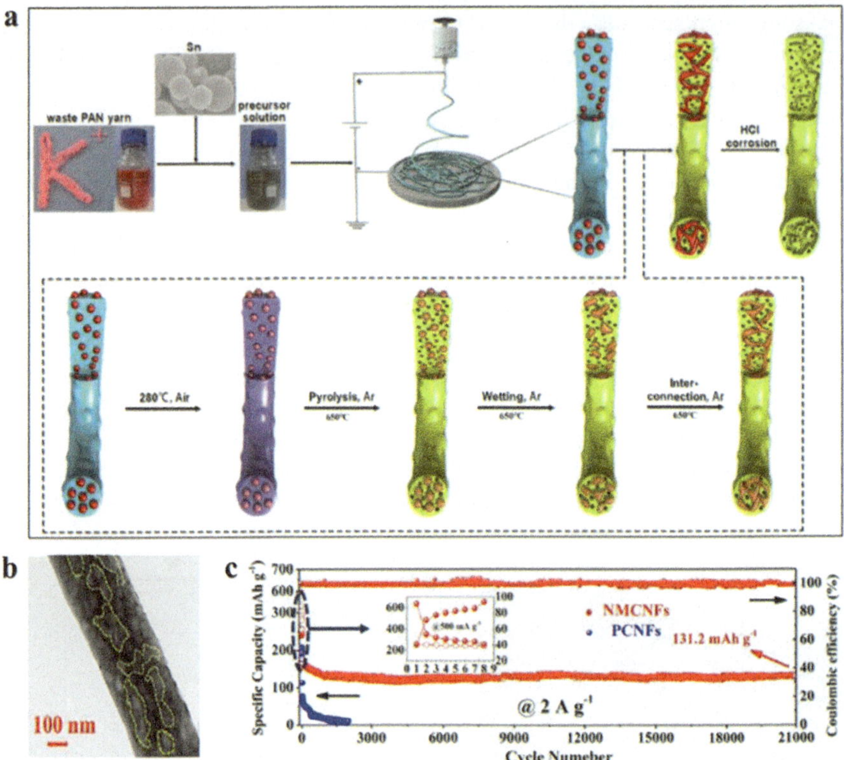

Fig. 4.6 **a** Schematic illustration of the fabrication procedure of NMCNFs. **b** HRTEM images of NMCNFs. **c** Ultra-long cycle performances and Coulombic efficiencies at a high current of 2 A g^{-1}. Reproduced from Ref. [43]. Copyright 2020, Elsevier

The corresponding TEM images further confirmed that as-made NMCNFs possessed average 300 nm diameter and high magnified view of TEM image exhibited lots of internally connected macropores existed inside of carbon nanofibers see Fig. 4.6b. Consequent long-term cycling of as prepared NMCNF anode discharge capacity of 185 mAh g^{-1} when operated at a high-rate of 2 A g^{-1} after activation of first seven cycles and maintained 129 mAh g^{-1} at 1000th cycle. During the following depotassiation/potassiation process, such an electrode shown unprecedented cycling stability, rendering a reversible capacity of 131.2 mAh g^{-1} after ultra-long cycling of 21,000 cycles, accompanying with 100% capacity retention Fig. 4.6c. The excellent performance attributes to the nominal interlayer spacing of NMCNFs is much larger, carbon nanofibers own the extremely stable porous structure and could accommodate the volume expansion. In supplement, the lots of defects and edges, highly pyridinic and pyrrolic N-doping could also provide enough active sites and upgrade the capability of K-ion adsorption [43].

4.5.2 2D/CNF Composites

Recent studies have focused on transition metal dichalcogenides (TMDs) like graphene analogue. TMDs are composed of group VI transition metals and chalcogens that received surging research interests as anode candidates for alkali-ion batteries [44]. Well known that the integration of metal chalcogenides with carbon materials provides not only a conducting network but also cushions large volume changes during charging and discharging. Miao et al. demonstrated a novel integrated 3D amorphous carbon-encapsulated CoS/nitrogen-doped carbon nanotube/CoS-coated carbon nanofiber (AC@CoS/NCNTs/CoS@CNFs) 3D network. The synthesis scheme described in Fig. 4.7a and their corresponding FE-SEM images shown in Fig. 4.7b. This structure reduces the ion diffusion path and improves the electronic conductivity and mechanical stability. The capacity is maintained at 130 mAh g^{-1} at 3.2 A g^{-1} after 600 cycles with excellent cycling stability. This superb performance was attributed to structural merits and capacitive effect.

The following equations were used to calculate the contributions from the diffusion and capacitive behaviours

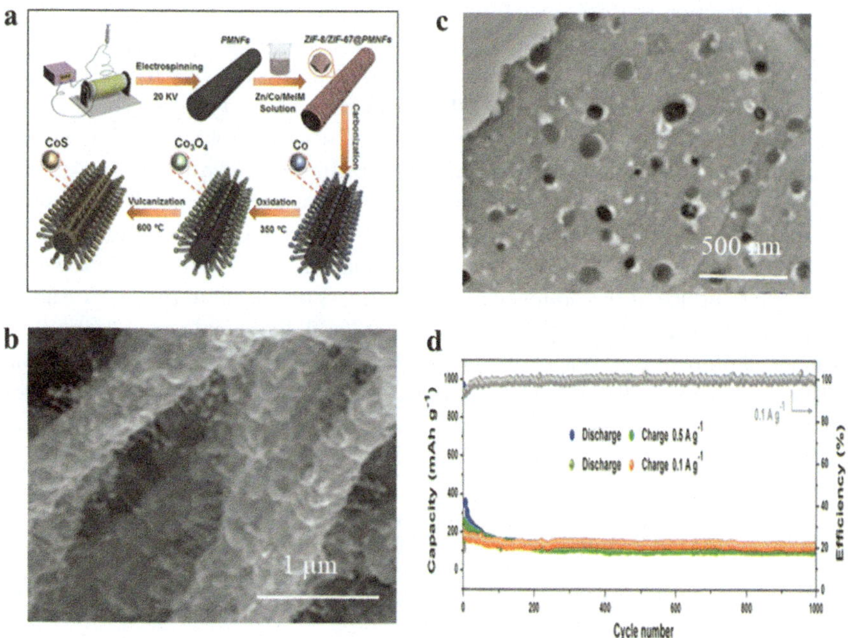

Fig. 4.7 **a** Schematic illustration of the preparation procedure for AC@CoS/NCNTs/CoS@CNFs. **b** FESEM images of AC@CoS/NCNTs/CoS@CNFs. Reproduced from Ref. [45]. Copyright 2019, The Royal Society of Chemistry. **c** TEM images of CoP@PPCS structure. **d** Cycling performance of potassium storage at current densities of 0.1 and 0.5 A g^{-1} and the CE at a current density of 0.1 A g^{-1} for CoP@PPCS structure. Reproduced from Ref. [46]. Copyright 2013 Wiley–VCH

$$i = k_1 v + k_2 v^{1/2} \tag{4.2}$$

$$i/v^{1/2} = k_1 v^{1/2} + k_2 \tag{4.3}$$

where, k_1 vis the capacitive component, and $k_2 v^{1/2}$ is the intercalation component [45]. In another literature, Bai et al. reported transition-metal phosphides (TMPs) based hybrid structures. A self-template and recrystallization-self-assembly strategy followed for the synthesis of core–shell-like cobalt phosphide (CoP) nanoparticles embedded into N and P co-doped porous carbon sheets (CoP@NPPCS) shown in Fig. 4.7c. As-obtained anode represents a reversible capacity of 127 and 114 mAh g^{-1} after 1000 cycles at 0.1 and 0.5 A g^{-1}, respectively, without detectable capacity decay [46] presented in Fig. 4.7d.

Among many prospective anodes, metal sulfides are being alternative anode to replace commercial anode for KIBs due to the higher electrical conductivity and other unique properties. Usually, ternary sulfides possess better electrochemical performance than binary sulfides because of their more abundant reaction sites and more reliable electronic conduction. Yet, due to the K ionic radius is greater than that of lithium and sodium, the volume structure of materials have changed during repeated insertion/extraction process, which will further influence the diffusion kinetics of K^+ and degrade the cycle stability of the material. Hence enhancing the diffusion kinetics of K^+ in ternary sulfides material for KIBs is necessary. Very recently, Zhang et al. designed novel NiCo$_2$S$_4$@N-HCNFs in which tapered rod-shaped nanomaterial NiCo$_2$S$_4$ and nitrogen-doped hollow carbon nanofibers. Figure 4.8a reveals the fabrication method in which involves two-step carbonization and hydrothermal processes, where the HCNFs are designed to synthesize by dual nozzle coaxial electrospinning of PMMA and PAN precursor solutions. When enduring a high-temperature hydrothermal reaction, a uniform tapered rod-shaped NiCo$_2$S$_4$ material grown-up on the outer surface of the hollow carbon nanofibers displayed in Fig. 4.8b. Subsequently, the cycling performance was conducted for all the sample to asses the electrochemical properties. It is apparat that the NiCo$_2$S$_4$@N-HCNFs electrode material possess a capacity of 701.2 mAh g^{-1} in the initial cycle. As the number of cycles increases, the CE exceeds 98%, which attribute that novel structural and excellent reaction mechanism. The reaction mechanism as follows from the CV curves [47]

$$NiCo_2S_4 + 8K^+ + 8e^- \leftrightarrow Ni + 2Co + 4K_2S \tag{4.4}$$

$$Ni + K_2S \leftrightarrow NiS + 2K \tag{4.5}$$

$$Co + K_2S \leftrightarrow CoS + CoS + 2K \tag{4.6}$$

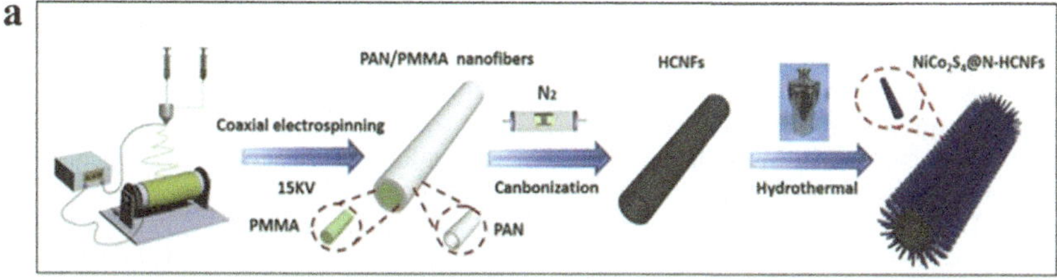

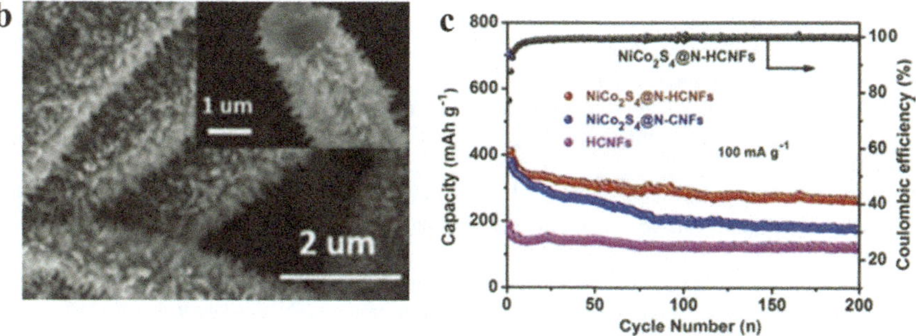

Fig. 4.8   **a** Schematic illustration of the preparation process of $NiCo_2S_4$ @N-HCNFs. **b** SEM image of $NiCo_2S_4$ @N-HCNFs. **c** Comparison of cycling performance at 100 mA g⁻¹. Reproduced from Ref. [47]. Copyright 2020 Elsevier

References

1. X. Bie, K. Kubota, T. Hosaka, K. Chihara, S. Komaba, A novel K-ion battery: hexacyanoferrate(II)/graphite cell. J. Mater. Chem. A **5**(9), 4325–4330 (2017)
2. C. Vaalma, G.A. Giffin, D. Buchholz, S. Passerini, Non-aqueous K-ion battery based on layered K0. 3MnO2 and hard carbon/carbon black. J. Electrochem. Soc. **163**(7), A1295–A1299 (2016)
3. N. Recham, G.l. Rousse, M.T. Sougrati, J.-N.l. Chotard, C. Frayret, S. Mariyappan, B.C. Melot, J.-C. Jumas, J.-M. Tarascon, Preparation and characterization of a stable FeSO4F-based framework for alkali ion insertion electrodes. Chem. Mater. **24**(22), 4363–4370 (2012)
4. V. Mathew, S. Kim, J. Kang, J. Gim, J. Song, J.P. Baboo, W. Park, D. Ahn, J. Han, L. Gu, Amorphous iron phosphate: potential host for various charge carrier ions. NPG Asia Mater. **6**(10), e138 (2014)
5. Y. Chen, W. Luo, M. Carter, L. Zhou, J. Dai, K. Fu, S. Lacey, T. Li, J. Wan, X. Han, Organic electrode for non-aqueous potassium-ion batteries. Nano Energy **18**, 205–211 (2015)
6. S. Komaba, T. Hasegawa, M. Dahbi, K. Kubota, Potassium intercalation into graphite to realize high-voltage/high-power potassium-ion batteries and potassium-ion capacitors. Electrochem. Commun. **60**, 172–175 (2015)
7. W. Luo, J. Wan, B. Ozdemir, W. Bao, Y. Chen, J. Dai, H. Lin, Y. Xu, F. Gu, V. Barone, Potassium ion batteries with graphitic materials. Nano Lett. **15**(11), 7671–7677 (2015)
8. D. Liu, Z. Yang, W. Li, Electrochemical Behavior of Graphite in KF–AlF3-Based Melt with Low Cryolite Ratio. J. Electrochem. Soc. **157**(7), D417–D421 (2010)
9. Z. Jian, W. Luo, X. Ji, Carbon electrodes for K-ion batteries. J. Am. Chem. Soc. **137**(36), 11566–11569 (2015)
10. I. Sultana, M.M. Rahman, Y. Chen, A.M. Glushenkov, Potassium-ion battery anode materials operating through the alloying-dealloying reaction mechanism. Adv. Func. Mater. **28**(5), 1703857 (2018)

11. V. Gabaudan, R. Berthelot, L. Stievano, L. Monconduit, Inside the alloy mechanism of Sb and Bi electrodes for K-Ion batteries. J. Phys. Chem. C **122**(32), 18266–18273 (2018)
12. T.T. Tran, M. Obrovac, Alloy negative electrodes for high energy density metal-ion cells. J. Electrochem. Soc. **158**(12), A1411–A1416 (2011)
13. X. Ren, Q. Zhao, W.D. McCulloch, Y. Wu, MoS 2 as a long-life host material for potassium ion intercalation. Nano Res. **10**(4), 1313–1321 (2017)
14. J. Ge, L. Fan, J. Wang, Q. Zhang, Z. Liu, E. Zhang, Q. Liu, X. Yu, B. Lu, MoSe2/N-doped carbon as anodes for potassium-ion batteries. Adv. Energy Mater. **8**(29), 1801477 (2018)
15. Q. Yu, J. Hu, C. Qian, Y. Gao, W.A. Wang, G. Yin, CoS/N-doped carbon core/shell nanocrystals as an anode material for potassium-ion storage. J. Solid State Electrochem. **23**(1), 27–32 (2019)
16. L. Fang, J. Xu, S. Sun, B. Lin, Q. Guo, D. Luo, H. Xia, Few-layered tin sulfide nanosheets supported on reduced graphene oxide as a high-performance anode for potassium-ion batteries. Small **15**(10), 1804806 (2019)
17. Y. Lu, J. Chen, Robust self-supported anode by integrating Sb2S3 nanoparticles with S, N-codoped graphene to enhance K-storage performance. Sci. China Chem. **60**(12), 1533–1539 (2017)
18. C. Chen, Z. Wang, B. Zhang, L. Miao, J. Cai, L. Peng, Y. Huang, J. Jiang, Y. Huang, L. Zhang, Nitrogen-rich hard carbon as a highly durable anode for high-power potassium-ion batteries. Energy Storage Mater. **8**, 161–168 (2017)
19. X. Wang, K. Han, D. Qin, Q. Li, C. Wang, C. Niu, L. Mai, Polycrystalline soft carbon semi-hollow microrods as anode for advanced K-ion full batteries. Nanoscale **9**(46), 18216–18222 (2017)
20. K. Share, A.P. Cohn, R.E. Carter, C.L. Pint, Mechanism of potassium ion intercalation staging in few layered graphene from in situ Raman spectroscopy. Nanoscale **8**(36), 16435–16439 (2016)
21. W. Cao, E. Zhang, J. Wang, Z. Liu, J. Ge, X. Yu, H. Yang, B. Lu, Potato derived biomass porous carbon as anode for potassium ion batteries. Electrochim. Acta **293**, 364–370 (2019)
22. Z. Tai, Q. Zhang, Y. Liu, H. Liu, S. Dou, Activated carbon from the graphite with increased rate capability for the potassium ion battery. Carbon **123**, 54–61 (2017)
23. X. Zhao, P. Xiong, J. Meng, Y. Liang, J. Wang, Y. Xu, High rate and long cycle life porous carbon nanofiber paper anodes for potassium-ion batteries. J. Mater. Chem. A **5**(36), 19237–19244 (2017)
24. R.A. Adams, J.-M. Syu, Y. Zhao, C.-T. Lo, A. Varma, V.G. Pol, Binder-free N-and O-rich carbon nanofiber anodes for long cycle life K-ion batteries. ACS Appl. Mater. Interfaces. **9**(21), 17872–17881 (2017)
25. B. Kishore, G. Venkatesh, N. Munichandraiah, K2Ti4O9: a promising anode material for potassium ion batteries. J. Electrochem. Soc. **163**(13), A2551 (2016)
26. Y. Xie, Y. Dall'Agnese, M. Naguib, Y. Gogotsi, M.W. Barsoum, H.L. Zhuang, P.R. Kent, Prediction and characterization of MXene nanosheet anodes for non-lithium-ion batteries. ACS Nano **8**(9), 9606–9615 (2014)
27. X. Ren, Q. Zhao, W. McCulloch, Y. Wu. Nano Res. **10**, 1313 2017; (b) Y, Wang, J. Wu, Y. Tang, X. Lu, C. Yang, M. Qin, F. Huang, X. Li, X. Zhang. ACS Appl. Mater. Interfaces **4**, 4246 (2012)
28. I. Sultana, M.M. Rahman, S. Mateti, V.G. Ahmadabadi, A.M. Glushenkov, Y. Chen, K-ion and Na-ion storage performances of Co3 O4–Fe2O3 nanoparticle-decorated super P carbon black prepared by a ball milling process. Nanoscale **9**(10), 3646–3654 (2017)
29. V. Lakshmi, Y. Chen, A.A. Mikhaylov, A.G. Medvedev, I. Sultana, M.M. Rahman, O. Lev, P.V. Prikhodchenko, A.M. Glushenkov, Nanocrystalline SnS2 coated onto reduced graphene oxide: demonstrating the feasibility of a non-graphitic anode with sulfide chemistry for potassium-ion batteries. Chem. Commun. **53**(59), 8272–8275 (2017)
30. H. Wang, X. Wu, X. Qi, W. Zhao, Z. Ju, Sb nanoparticles encapsulated in 3D porous carbon as anode material for lithium-ion and potassium-ion batteries. Mater. Res. Bull. **103**, 32–37 (2018)

31. I. Sultana, T. Ramireddy, M.M. Rahman, Y. Chen, A.M. Glushenkov, Tin-based composite anodes for potassium-ion batteries. Chem. Commun. **52**(59), 9279–9282 (2016)

32. X. Zhao, Y. Tang, C. Ni, J. Wang, A. Star, Y. Xu, Free-Standing Nitrogen-Doped Cup-Stacked Carbon Nanotube Mats for Potassium-Ion Battery Anodes. ACS Appl. Energy Mater. **1**(4), 1703–1707 (2018)

33. W. Yang, J. Zhou, S. Wang, W. Zhang, Z. Wang, F. Lv, K. Wang, Q. Sun, S. Guo, Free-standing film made by necklace-like N-doped hollow carbon with hierarchical pores for high-performance potassium-ion storage. Energy Environ. Sci. (2019)

34. T. Jin, H. Li, Y. Li, L. Jiao, J. Chen, Intercalation pseudocapacitance in flexible and self-standing V2O3 porous nanofibers for high-rate and ultra-stable K ion storage. Nano Energy **50**, 462–467 (2018)

35. Y. Wu, S. Hu, R. Xu, J. Wang, Z. Peng, Q. Zhang, Y. Yu, Boosting potassium-ion battery performance by encapsulating red phosphorus in free-standing nitrogen-doped porous hollow carbon nanofibers. Nano Lett. **19**(2), 1351–1358 (2019)

36. S. Liu, Z. Wang, C. Yu, H.B. Wu, G. Wang, Q. Dong, J. Qiu, A. Eychmüller, X.W. Lou, A flexible TiO2 (B)-based battery electrode with superior power rate and ultralong cycle life. Adv. Mater. **25**(25), 3462–3467 (2013)

37. Y. Zhu, X. Han, Y. Xu, Y. Liu, S. Zheng, K. Xu, L. Hu, C. Wang, Electrospun Sb/C fibers for a stable and fast sodium-ion battery anode. ACS Nano **7**(7), 6378–6386 (2013)

38. Y.h. Zhu, Y.b. Yin, X. Yang, T. Sun, S. Wang, Y.s. Jiang, J.m. Yan, X.b. Zhang, Transformation of rusty stainless-steel meshes into stable, low-cost, and binder-free cathodes for high-performance potassium-ion batteries. Angewandte Chem. Int. Ed. **56**(27), 7881–7885 (2017)

39. M. Mao, C. Cui, M. Wu, M. Zhang, T. Gao, X. Fan, J. Chen, T. Wang, J. Ma, C. Wang, Flexible ReS2 nanosheets/N-doped carbon nanofibers-based paper as a universal anode for alkali (Li, Na, K) ion battery. Nano Energy **45**, 346–352 (2018)

40. M. Inagaki, Y. Yang, F. Kang, Carbon nanofibers prepared via electrospinning. Adv. Mater. **24**(19), 2547–2566 (2012)

41. Y. Liu, F. Fan, J. Wang, Y. Liu, H. Chen, K.L. Jungjohann, Y. Xu, Y. Zhu, D. Bigio, T. Zhu, In situ transmission electron microscopy study of electrochemical sodiation and potassiation of carbon nanofibers. Nano Lett. **14**(6), 3445–3452 (2014)

42. C. Lv, W. Xu, H. Liu, L. Zhang, S. Chen, X. Yang, X. Xu, D. Yang, 3D Sulfur and nitrogen codoped carbon nanofiber aerogels with optimized electronic structure and enlarged interlayer spacing boost potassium-ion storage. Small **15**(23), 1900816 (2019)

43. S. Zhang, Z. Xu, H. Duan, A. Xu, Q. Xia, Y. Yan, S. Wu, N-doped carbon nanofibers with internal cross-linked multiple pores for both ultra-long cycling life and high capacity in highly durable K-ion battery anodes. Electrochim. Acta **337**, 135767 (2020)

44. S. Wu, Y. Du, S. Sun, Transition metal dichalcogenide based nanomaterials for rechargeable batteries. Chem. Eng. J. **307**, 189–207 (2017)

45. W. Miao, Y. Zhang, H. Li, Z. Zhang, L. Li, Z. Yu, W. Zhang, ZIF-8/ZIF-67-derived 3D amorphous carbon-encapsulated CoS/NCNTs supported on CoS-coated carbon nanofibers as an advanced potassium-ion battery anode. J. Mater. Chem. A **7**(10), 5504–5512 (2019)

46. J. Bai, B. Xi, H. Mao, Y. Lin, X. Ma, J. Feng, S. Xiong, One-step construction of N, P-codoped porous carbon sheets/CoP hybrids with enhanced lithium and potassium storage. Adv. Mater. **30**(35), 1802310 (2018)

47. W. Zhang, J. Chen, Y. Liu, S. Liu, X. Li, K. Yang, L. Li, Decoration of hollow nitrogen-doped carbon nanofibers with tapered rod-shaped NiCo2S4 as a 3D structural high-rate and long-lifespan self-supported anode material for potassium-ion batteries. J. Alloys Compd. 153631 (2020)

Chapter 5
Electrospinning of Nanofibers for Li–S Battery

Abstract Energy generation and storage are vital to research fields where the requests for increased energy devices and the necessity for greener energy supplies are rising. The Li–S battery exhibits excellent potential and has attracted the attention of battery developers in large scale production in recent years on account of its low cost, theoretically large specific capacity, probably high specific energy, and its eco-friendly footprint. Several concerns associated in the development of Li–S batteries, particularly shuttle the phenomenon and hence many efforts and improvements have been made in the past decades. Noticeably, we review the fundamental research and technological development of electrospun nanofiber materials for Li–S cells, including their handling methods, constructions, morphology engineering, and electrochemical performance.

Keywords Li–S battery · Fiber-based electrode · Polysulfides · Electrochemical characteristics

5.1 Li–S Battery Working Principle and Cell Structure

Attainment apart the hope of rechargeable lithium batteries needs a search of new chemistry, particularly electrochemistry, and new materials. Non-lithiated cathode materials have shown higher specific capacities than lithiated cathode materials also deliver boosted safety. Rechargeable lithium-sulfur (Li–S) is decidedly capable lithium rechargeable battery since sulfur has the highest theoretical capacities. According to the reaction mechanisms with metal lithium to form Li_2S, it has a theoretical specific capacity of 1675 mAh g^{-1} [1–3]. The Li–S battery composed of a sulfur composite cathode, a polymer or liquid electrolyte, and a lithium anode. A schematic outline of a Li–S battery illustrated in Fig. 5.1.

Generally, Li reacts with sulfur (S_8) to produce lithium polysulphides with a formula of Li_2S_n. Long-chain polysulphides are produced first such as Li_2S_8 and Li_2S_6, which are condensed during the further reduction of sulfur. The overall reaction can be written as follows [3]

© The Editor(s) (if applicable) and The Author(s), under exclusive license   101
to Springer Nature Singapore Pte Ltd. 2020
S. Peng and P. R. Ilango, *Electrospinning of Nanofibers for Battery Applications*,
https://doi.org/10.1007/978-981-15-1428-9_5

Fig. 5.1 A schematic outline of a Li–S battery. Reproduced from Ref. [3]. Copyright, 2013 The Electrochemical Society

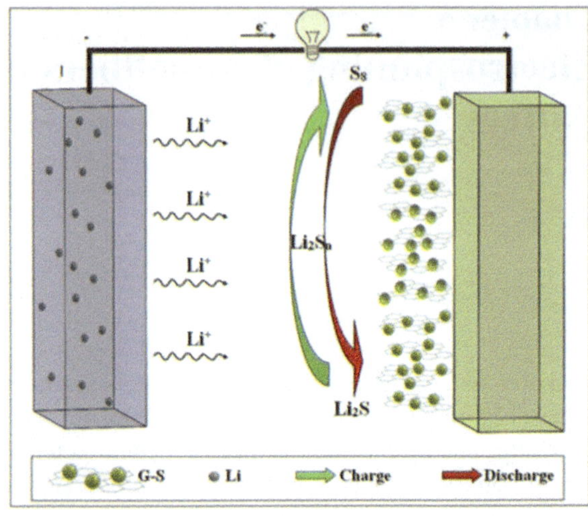

$$16\text{Li} + \text{S}_8 \rightarrow 8\text{Li}_2\text{S} \qquad\qquad (5.1)$$

5.2   Perspectives on Material Development

The recipe Li–S is one of the most promising chemistries for the next generation of lithium batteries, with high specific energy (2500 Wh kg^{-1}) and high theoretical capacity that are noticeably higher than LIBs. The utmost vital challenges are the low conductivity of sulfur and of its solid reduction products, Li$_2$S$_2$ and Li$_2$S, and the high solubility of intermediate products formed in typically used liquid-electrolyte when sulfur is step by step lithiated [4]. The soluble intermediates will deposit on both anode and cathode that lead to unsatisfactory Li–S battery performance. Over the past decade, considerable growth has been received to minimize the drawbacks, resulting in substantial advances in terms of the specific capacity, cycling stability and calendar life of Li–S batteries. For example, carbon has been used as a host for sulfur which drawn better cycle life and coulombic efficiency [5, 6].

5.2.1   Cathode

The elemental sulfur is commonly used as cathode material in Li–S batteries, among which cyclo-S$_8$ is the most thermodynamically stable allotrope under ambient conditions. Typical sulfur electrode needs the ensuing properties of a closed structure for efficient polysulphide restraint, an inadequate surface area for sulfur electrolyte contact and satisfactory room to accommodate sulfur volumetric growth [7]. The main challenges of the cathode are its miserable cycle life, low coulombic efficiency, and relatively large self-discharge rate due to the non-conductivity of S and dissolved polysulfide products from the cathode area. To overcome these challenges, many methods have been subjected. For example, conductive materials like carbon used to improve the conductivity of the cathode by following various physical or chemical adsorption to block polysulfides. Numerous research efforts have concentrated on the progress of carbon/sulfur nanocomposites, in which sulfur particles were embedded in the nanopores of the conductive carbon matrix resulting increase of both electrical and ionic conductivity of the sulfur cathode while at the same time suppressing the polysulphide shuttle principle [8, 9]. Among several methods, electrospinning skill displays several benefits such as simple, versatile, and low cost. When use nanofibers or nanofibers based cathode composite, the Li–S delivers great electrochemical performance due to exceptional features comprising excellent surface area and flexibility [10, 11]. According to the previous literature, the cathode material can be classified as follows

1. Non-doped carbon cathode material
2. The doped carbon cathode material
3. Inorganic cathode material
4. Polymeric cathode material.

In the recent year's great development in Li–S cells have been added by electrospun fibers based on their innovative physicochemical compensations, multitudinous tasks such as cycling stability, energy density, and safety are still tackled for its viable application.

5.2.2 Anode

Li-metal is served as an anode in rechargeable Li–S batteries. The usage of Li as the anode in Li–S batteries continues a key topic owing to safety matters resulting from the establishment of Li dendrites throughout cycling, that could enter the separator and tip to thermal runaway. Moreover, the unremitting reaction of the soluble polysulfide to the Li anode primes to noteworthy self-discharge and the deposition of solid Li_2S_2 and Li_2S on the cathode, which outcomes inactive mass loss and capacity declining. Alternatively, to evade this safety difficult in the Li–S system is to practice a high-capacity anode material other than Li metal. Earlier, Yang et al. have explored Li-metal-free Li–S battery involving a silicon nanowire anode and a Li_2S/mesoporous carbon composite cathode. This new battery yields theoretical specific energy of 1550 Wh kg^{-1}, which is four times that of the theoretical specific energy has been stated [12]. Afterwards, Li metal has been replaced by commercial graphite [13], Sn–C–Li [14], Li–Al alloy [15] and Lithium-rich $Li_{2.6}BMg_{0.05}$ [16] Li–Si [17] alloy by several groups and stated that the new anode could inhibits the dendrites formation and suppress the soluble polysulphides and maintains the stable cycle life. Further many attempts have been devoted recently, including a formation of

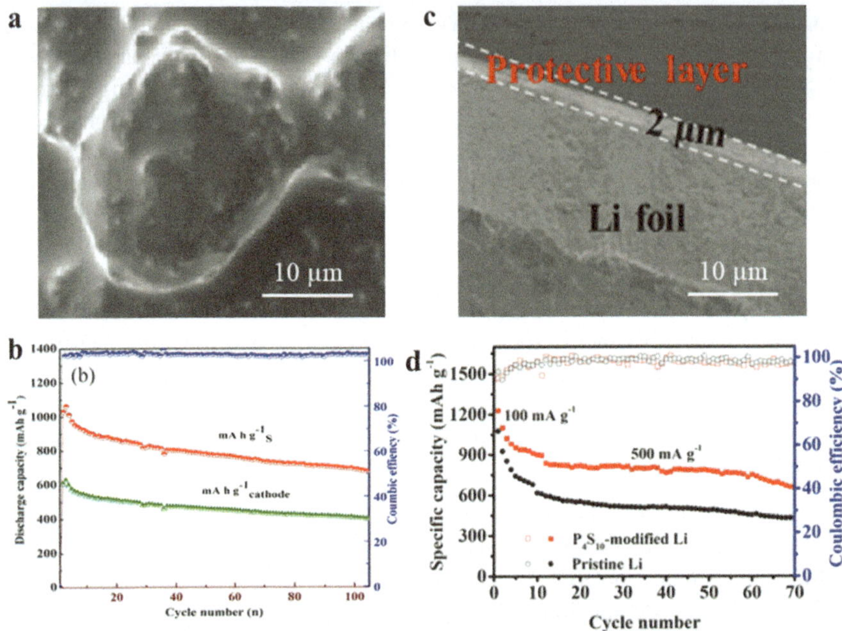

Fig. 5.2 **a** SEM images of the surface of graphite electrode in lithiated graphite-sulfur (Li–S) full cell with 5 M LiTFSI-DME/DOL electrolytes after 50 cycles. **b** Cyclic performances of lithiated graphite-sulfur full cell with 5 M LiTFSI-DME/DOL at a rate of 0.1 °C (1675 mA g^{-1} based on sulfur cathode). Reproduced from Ref. [18]. Copyright 2017, Elsevier. **c** The cross-sectional SEM of P_4S_{10}-modified Li foil. **d** Cycling performance of Li–S batteries with modified Li and pristine Li anodes at 500 mA g^{-1}. Reproduced from Ref. [19]. Copyright 2019, Elsevier

stable surface film by using super concentrated ether (LiTFSI-DME/DOL) electrolyte on graphite electrode. Figure 5.2a reveals the SEM images of the surface of graphite electrode in lithiated graphite-sulfur (Li–S) full cell with 5 M LiTFSI-DME/DOL electrolytes after 50 cycles and resulting fairly high specific capacity of 1031 mAh g^{-1} $_{sulfur}$ at 0.1 °C (1 °C = 1675 mA g^{-1}), and a reversible capacity of 686 mAh g^{-1} $_{sulfur}$ can still be conserved after 105 cycles see in Fig. 5.2b [18]. Very recently, Le et al. introduced P_4S_{10} modified lithium anode for enhanced performance of Li–S batteries. After modification with P_4S_{10} solution, the Li foil demonstrated a quite smooth and undeviating surface with some nanoscale bumps and the cross-sectional SEM image of the modified Li is presented in Fig. 5.2c from which the thickness of the protective layer was assessed to be 2 μm. The battery test with pristine Li anodes validated an initial specific capacity of 1077.6 mAh g^{-1} and quickly decreased to 427.1 mAh g^{-1} after 70 cycles. Differently, batteries test with modified Li anodes exhibited improved cycling performance, continuing stable discharge capacity of 699.5 mAh g^{-1} after 70 cycles with the average CE of 97.6% shown in Fig. 5.2d. Moreover, creating $Li_xP_yS_z$ based shielding layer is an operational approach to steady

Li metal anodes and a capable tactic to encourage the real-world use of advanced Li-metal batteries [19].

5.2.3 Binder-Free Electrodes

With the quick expansion of flexible screens and circuits, the flexibility of batteries is fetching important. While flexible LIBs have been extensively described, they cannot content gradually high energy request. Li–S batteries own high theoretical capacity and natural abundance [20]. More prominently, the theoretical energy densities of Li–S batteries can reach 2500 Wh kg^{-1}, which is ca. five times higher than that of commercial LIBs [21]. Thus, Li–S batteries have been reflected in a favourable aspirant for next-generation flexible energy storage devices. Although the prominent theoretical benefits of Li–S batteries, many difficulties still need to be solved before the application of flexible Li–S batteries. Among them, the fabrication of flexible electrodes is a tough task. Binder free electrodes show remarkable performance in flexible Li–S batteries [22, 23]. Many efforts have been made recently by many research groups [24, 25]. Expressively, electrospinning technique is a vital process to prepare binder-free electrodes, and it had been used as the cathode in Li–S system [26].

5.2.4 Binder-Free Cathodes

To address the Li–S problems, earlier Zu et al. reported Li/dissolved polysulfide cells with a binder-free carbon nanofiber (CNF) paper electrode that grants a low trade cost benefit over the MWCNT electrodes. As prepared CNF SEM image shown in Fig. 5.3a which has micron-sized interspaces that are at least an order of degree larger than those of the MWCNT paper. Furthermore, the CNF has a diameter of 200 nm, which is also an order of degree superior to that of the carbon nanotube (20 nm). The great interspaces within the CNF paper electrode could host more cycled products, that is sustained by the high porosity (97%) of the CNF paper electrode, expected based on the density of graphite, the volume, and weight of the circular electrode disk. Besides, the thick nanofibers can maintain a healthy electrode structure. To get insights, their electrochemical performance was examined. Figure 5.3b represents the cycle performance and coulombic efficiency of the CNF paper/dissolved polysulfide cell at C/5 rate between 2.6 and 1.8 V. Noted, that a high discharge capacity of 1139 mAh g^{-1} (based on the mass of sulfur in the catholyte) is acquired through the 2nd discharge (the 1st discharge only has the low-voltage plateau), confirming to high consumption of active polysulfide species [24]. Later, Lu et al. proposed a novel binder-free cathode (S-FCNF) in which effectively functionalized carbon nanofibers coating with sulfur. The TEM images of S-FCNF shown in Fig. 5.3c. A uniform

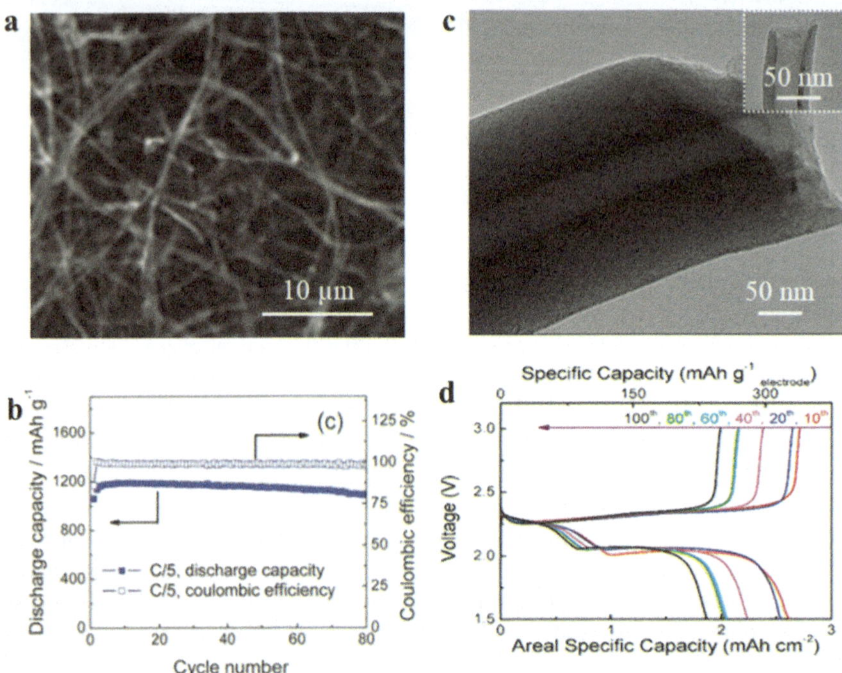

Fig. 5.3 **a** SEM images of CNF paper. **b** Cycle performance and Coulombic efficiency at C/5 rate between 2.6 and 1.8 V. Reproduced from Ref. [24]. Copyright 2013, The Royal Society of Chemistry. **c** Typical TEM image of S-FCNFs with a uniform sulfur coating (inset image is FCNFs). **d** Charge–discharge profiles of S-FCNFs electrode at different cycle numbers is labelled. Reproduced from Ref. [25]. Copyright 2014, The Royal Society of Chemistry

tubular layer of sulfur was coated on the surface of FCNFs. It formed nano microsphere structure with a thickness of 45 nm while the uncoated FCNFs TEM image illustrated in Fig. 5.3c (inset). This structure could adopt huge volume expansion and absorb high polysulfides and helps to improve the areal specific discharge capacity of 4.49 mAh cm^{-2} (561 mAh $g_{electrode}^{-1}$) initial discharging process [25] as presented in Fig. 5.3d.

Afterwards, several conducting polymers of high conductivity and thermal stability have been discovered as conductive materials to substitute carbon material also used as cathode materials for Li–S batteries [27, 28]. Recently, Zhang et al. introduced D-Sulfur/SWCNT in which PEDOT-coated diamond-shaped sulfur (D-sulfur)/single-walled carbon nanotube (SWCNT) composite cathode. The preparation of diamond sulfur was based on a simple reaction as follows

$$Na_2S_2O_3 \rightarrow S \downarrow + SO_2 \uparrow + NaCl + H_2O \tag{5.2}$$

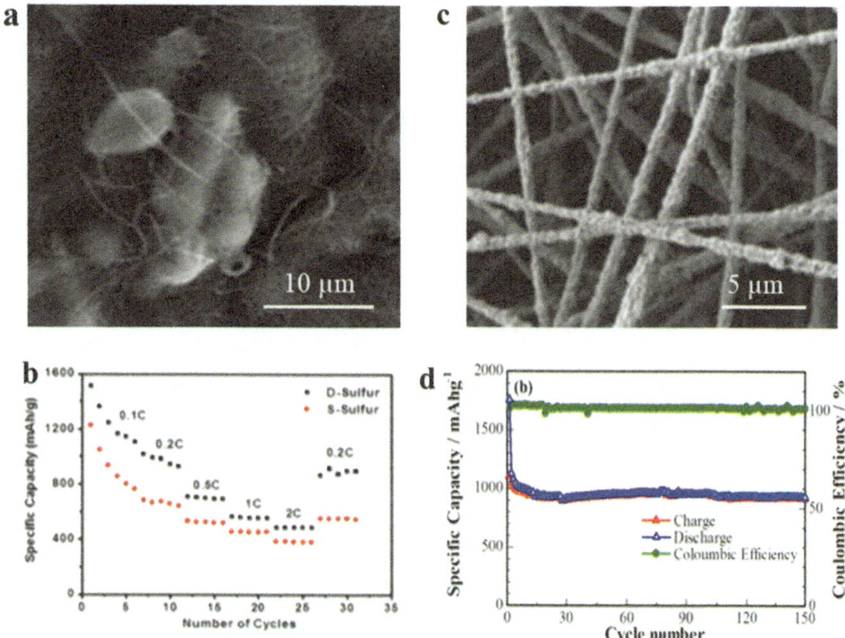

Fig. 5.4 **a** SEM images of D-sulfur/SWCNT composite cathode. **b** Rate capability of the D-sulfur/SWCNT and S-sulfur/SWCNT composite cathodes. Reproduced from Ref. [22]. Copyright 2017, The Royal Society of Chemistry. **c** SEM images of S/DPAN/KB composite. **d** Cycle performance S/DPAN/KB composite cathode nanofibers at 0.1 °C Reproduced from Ref [26]. Copyright, The Author(s) 2019

PVP molecules played a vital role, which performed as a capping agent to switch particle growth and a soft structural template of sulfur. The SEM images of D-sulfur/SWCNT composite is demonstrated in Fig. 5.4a in which the sulfur particles revealed a distinct diamond shape, in which the short diagonal of D-sulfur was calculated to be 450 nm. From the rate capability test displayed in Fig. 5.4b, they conclude that D-sulfur/SWCNT was higher than that of S-sulfur/SWCNT due to encapsulation by PEDOT [22]. Another exciting research is combining conductive polymers with carbon material. For instance, Kalybekkyzy et al. fabricated S/DPAN/KB binder-free cathode in which polyacrylonitrile nanofibers filled with Sulfur/Ketjen Black particles. The SEM image of S/PAN/KB nanofibers composite fibres revealed a rough surface with the diameter ranging from 300 to 600 nm (see in Fig. 5.4a). The cycling performance of as-prepared binder-free S/DPAN/KB composite delivers stable cycling behaviour with less capacity fading about 0.017% which ascribes to the shrinking of the polysulfides shuttle effect by the molecular level binding of sulfur with DPAN [26]. Not long ago, pyrite (cubic-FeS$_2$) has recaptured devotion as a capable cathode material for LIBs because of its high theoretical energy density ($\sim$1313 Wh/kg based on the reaction of FeS$_2$ + 4Li = Fe + 2Li$_2$S), low cost, earth abundance, and nontoxicity. Hence, *Y. Zhu* et al. developed FeS$_2$@carbon fiber as the binder-free electrode. This electrode not only increases the reaction kinetics through nanosized FeS_2 and conductive network but also evade the effect of the current collector and/or other components (e.g., binder) on the redox mechanisms and cycling stability of Li-FeS_2. Based on the early literature survey, the first lithiation of FeS_2 ensued via an intermediate Li_2FeS_2 by subsequent a two-step reduction principle [29, 30]

$$FeS_2 + 2Li^+ + 2e^- \rightarrow Li_2FeS_2 \tag{5.3}$$

$$Li_2FeS_2 + 2Li^+ + 2e^- \rightarrow 2Li_2S + Fe \tag{5.4}$$

Subsequently, very small Fe particles and Li_2S were generated in the final reduction products which have been analyzed by many researchers [31, 32].

5.3 Free-Standing Electrodes

5.3.1 Free-Standing Cathodes

The free-standing electrode will be another brilliant strategy to improve the conductivity of and inhibiting the solubility of lithium polysulfides. Modern investigation designates that transition metals such as Fe [33], Ni [34] Cu [35] can react with S and form transition metal sulfides that can be used as cathode materials for Li–S batteries with remarkable performances. For example, Zeng et al. free-standing copper-immobilized sulfur-porous carbon nanofiber (S@PCNFs-Cu) to overcome the so-called drawbacks of Li–S batteries. Figure 5.5a represents a preparation scheme for the S@PCNFs-Cu electrode. Herein, they have proposed that micropores and mesopores of PCNF hosts deliver enough space to receive the volume alter of S and polysulfide. Moreover, the voltage profiles of the S@PCNFs-Cu in promise with the peaks in the CV with the initial discharge and charge capacities of 1728 and 892 mA h g^{-1}, respectively, consistent to an initial coulombic efficiency (ICE) of 51.6% shown in Fig. 5.5b [36]. Later, many carbons based freestanding electrode with metal impregnation has been reported to control the shuttle phenomena of polysulfides such as $S_{1-x}Se_x$@PCNFs and $S_{0.6}Se_{0.4}$@CNFs due to the S-Se bonding and porous structure [37, 38]. Also, freestanding metal oxide has been proposed, for example, MoO_2–CNFs [39], Al_2O_3–C [40], Ti_4O_7–C [41], and TiO–CNF [42] have demonstrated great interest in solving the polysulfide diffusion conundrum due to polar unites can adsorb the migrating polysulfides by the chemical interface. Recently, most of the novel interlayers based on various metal oxides/carbon materials have been applied to Li–S batteries. Among them, TiO_2 based composite has significant notice owing to its low cost, harmless, and facile synthesis process. For example, Singh et al. proposed freestanding TiO/CNF-S cathodes by electrospinning process. The whole preparation strategy outlined by the scheme in Fig. 5.5c.

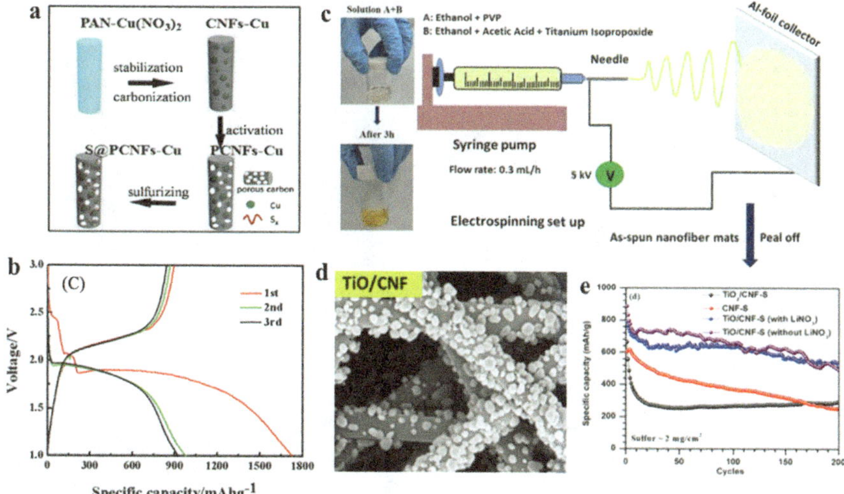

Fig. 5.5 **a** Schematic illustration of the synthesis process for the S@PCNFs-Cu electrode. **b** Voltage profiles of S@PCNFs-Cu. Reproduced from Ref. [36]. Copyright 2015, The Royal Society of Chemistry. **c** Schematic representation of TiO/CNF. **d** SEM images of freestanding TiO/CNF nanofiber mats. **e** Cyclic stability test at 0.5 °C rate for as prepared (with and without LiNO₃) cathodes over 200 cycles. Reproduced from Ref. [42]. Copyright 2018, American Chemical Society

Titanium isopropoxide (TiP)/poly-(vinylpyrrolidone) (PVP) in ethanol/acetic acid solvents used. The acquired as-spun nanofiber mats from this gel were then heat-treated at 600 and 950 °C to prepare TiO$_2$/CNF and TiO/CNF samples, respectively. As developed freestanding mat SEM image shown in Fig. 5.5d. It can be noted that as developed nanofibrous structures with an outer diameter of about 500–900 nm. The comparative cycling performance results suggest that the fabricated freestanding TiO/CNF-S cathodes exhibit high initial discharge capacities of ∼1080, ∼975, and ∼791 mAh g^{-1} at 0.1, 0.2, and 0.5 °C rates, respectively, with stable cycling performance over prolonging cycling [42] see in Fig. 5.5e. Moreover, XPS and Raman analysis have been used in this research for the first time to disclose the occurrence of strong Lewis acid–base interaction between TiO (3d^2) and S$_x$$^{2-}$ over the coordinate covalent Ti-S bond establishment. Excitingly, Li et al. approached novel ternary freestanding NiO/RGO-Sn composite and examined their immobilizing effect of NiO and Sn for sulfur species by using XPS studies and claimed that NiO/RGO-Sn hybrid film exhibits the outstanding electrochemical performance of Li–S battery which is ascribed to the hybrid structure contained Sn and NiO with RGO [43].

Besides, carbonaceous materials mainly include porous carbon [44], carbon nanotube [45], and graphene [46] have been established as freestanding to mitigate those drawbacks. Furthermore, 3D hierarchically porous electrode contained stringed N-doped hollow carbon spheres (SN-HCSs) loaded with sulfur element utilized as the freestanding cathode and the preparation scheme illustrated in Fig. 5.6a. The

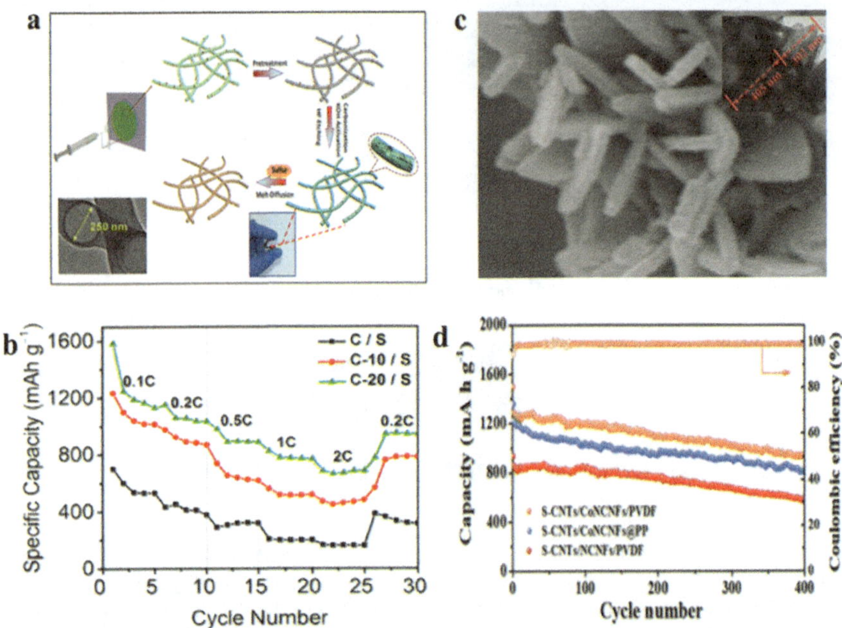

Fig. 5.6 **a** Scheme for fabrication process for the flexible 3D hierarchically porous sulfur composite electrode comprised of stringed N-doped hollow carbon spheres (inset: TEM image) **b** The rate capacities of C-20/S, C-10/S and C/S electrodes at different current densities 2C. Reproduced from Ref. [47]. Copyright 2017, Elsevier. **c** SEM images of CoNCNFs layer (inset: Cryo-TEM images of CoNCNFs). **d** Cycling performance of as-prepared freestanding membrane. Reproduced from Ref. [48]. Copyright 2019 Wiley–VCH

image in Fig. 5.6a (inset) reveals a TEM image of the C-20, which shows a multi-wall graphitized hollow carbon structure with the interplanar distance of 0.38 nm. The flexible free-standing C-20/S cathode also displays a much better rate capability than C-10/S and C/S (Fig. 5.6b). It delivers a reversible capacity as high as 1249.0, 1062.5, 892.6, 781.4, and 688.4 mAh g^{-1} when cycled at the current density of 0.1, 0.2, 0.5, 1, and 2 C. Once the current density is reverted to 0.2 C, the specific discharge capacity of the C-20/S cathode recovers to 954.7 mAh g^{-1}, close-fitting a strong tolerance for high current [47]. Very recently, high-flux, flexible, electrospun fibrous membranes are advanced, which succeed in incorporating three functional units (cathode, interlayer, and separator) into an efficient composite. As prepared three-in-one S-CNTs/CoNCNFs/PVDF membrane is believed towards negative interface effect. Figure 5.6c reveals SEM images of CoNCNFs layer and inset images shows Cryo-TEM images of CoNCNFs. The S-CNTs/CoNCNFs/PVDF membrane presented a high initial discharge capacity of 1501 mAh g^{-1} at 0.2 C, which was engaged at 933 mAh g^{-1} after 400 cycles [48]. It also produced an alleviated coulombic efficiency above 99% exposed in Fig. 5.6d. Meanwhile, there are many sulfide, selenide nitrates and carbides based materials reported such as MoS$_2$

[49], MoSe$_2$ [50], Mo$_2$C [51, 52], VN [53] Ti$_3$C$_2$ [54] and Fe$_3$C [55]. They are demonstrating the most excellent Li–S electrochemical performance.

5.4 Other Electrodes

5.4.1 Interlayer for Li–S Batteries

Besides, the introduction of the interlayer between the cathode and separator could also hugely expand Li–S battery characteristics [56, 57]. The electrospun nanofibers as interlayer is greatly strengthened the conductivity also noticeably destroy the transfer of polysulfides within the cathode side resulting in high capacity retention over the cycles [58]. By now, many electrospun CNFs-based interlayers have been evidenced to be appropriate interlayers for Li–S batteries with good conductivity and controlled porous structures since porous nature is very advantageous for hindering active S material and directorial Li growth [58, 59]. Previously, Manthiram's group suggested a novel cell configuration based on a carbon interlayer inserting between the cathode and the separator [23, 60, 61]. Porous carbon particles [60], carbon nanotubes [61], carbonized leaf/Kimwipes [62], reduced graphene oxide films [63], carbon nanofibers [44] have been validated to be suitable interlayers for Li–S batteries. Among them, electrospun nanofibers have merits of fine diameter, admirable conductivity, and unpaid thermal stability, which facilitate the electrolyte penetration [58, 64]. For example, Wang et al. prepared a flexible activated carbon nanofiber (ACNF) interlayer between the sulfur cathode and the separator via electrospinning polyimide and activation treatment. A model scheme of a Li–S battery with an ACNF paper as interlayer showed in Fig. 5.7a. The consistent pore size distribution is demonstrated in Fig. 5.7b, which represents a noteworthy comment of micropores and mesopores located in the range of 0.9–2 and 2.5–10 nm, respectively. Additionally, the mesopores are understood to be favourable for channelling dissolved polysulfide ions into the micropores. In Fig. 5.7b (inset), displays a general morphology of the as-fabricated ACNFs that is apparent and self-possessed of regular and randomly oriented nanofibers, founding an interconnected and porous network structure. The length of these nanofibers can extend a few hundreds of microns with the diameter ranging from 50 to 500 nm. The lightweight ACNF interlayer with the higher specific surface area could produce improved performance at a low mass ratio of ACNF/sulfur (0.4) with an initial specific capacity of 1224 mAh g^{-1} along with superb coulombic efficiency [65].

Consequently, PAN-NC interlayer developed in which polyacrylonitrile and N-doped carbon black fiber are directly coated S cathode through the electrospinning process. As fabricated, PAN-NC interlayer keeps porous fiber structure, that can offer outstanding electrolyte infiltration and flexibility to buffer the structure alter,

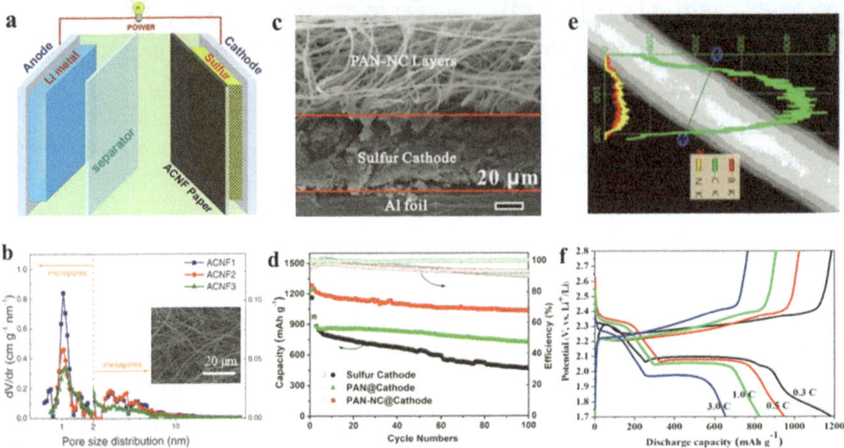

Fig. 5.7 **a** Schematic illustration of a Li–S cell configuration with an ACNF paper inserting between cathode and separator. **b** Pore size distribution of ACNF. A SEM image of the ACNF paper (inset). Reproduced from Ref. [65]. Copyright 2015, Elsevier. **c** Cross-sectional SEM image of the PAN-NC@cathode. **d** Cycling performances at 200 mA g^{-1} of multifunctional interlayer. Reproduced from Ref. [66]. Copyright 2017, American Chemical Society. **e** EDX line scan with element distributions on the cross-section of BNCNF. **f** Charge–discharge curves of the Li–S cells with BNCNF interlayer. Reproduced from Ref. [68]. Copyright 2018, Elsevier

ensuring fast Li$^+$ diffusion and durable bond in cycling. Also, the PAN-NC interlayer performances as a second current collector since of its good affinity for polysulfides and anticipated conductivity for electron transport. The thickness of the PAN-NC@cathode is analyzed by the cross-sectional image in see Fig. 5.7c. The PAN-NC layer with the thickness of ∼50 μm can be detected on the pure sulfur cathode of ∼60 μm. Further, they evaluated the cycling performances of the cathodes are compared at the current density of 200 mA g^{-1} demonstrated in Fig. 5.7d.

It can be noted that the higher discharge capacity is acquired for the PAN-NC@cathode, with a residual reversible capacity of 1029 mAh g^{-1} over 100 cycles than pure S cathode (467 mAh g^{-1}) that, implying the aid of the PAN-NC interlayer in suppressing polysulfide dissolution [66]. Interestingly, Melamine was endorsed as a nitrogen-rich amine element to prepare a modified polyacrylic acid (MPAA). The electrospun MPAA was carbonized into N-rich carbon nanofibers, which were used as cathode interlayers reveals the initial discharge capacity with two interlayers was 983 mAh g^{-1} and faded down to 651 mAh g^{-1} after 100 cycles with the coulombic efficiency of 95.4% at 0.1 °C rate [67]. Very recently, the shuttle effect of soluble lithium polysulfides (LiPSs) and their insoluble reduction products (Li$_2$S/Li$_2$S$_2$) have been controlled by using boron and nitrogen co-doped carbon nanofibers (BNCNF) as a facile electrospinning technique has produced interlayer with subsequent thermal treatment.

The doping contents of B and N elements on the cross-section of BNCNF are also tactic by the same intensity as shown in Fig. 5.7e. Additional approving the consistent supply of B and N elements through the CNF framework. To get insights, they examined discharge/charge profiles of Li–S battery with BNCNF interlayer at the current density from 0.3–3.0 °C display two typical discharge plateaus at 2.32 V and 2.10 V as exposed in Fig. 5.7f. which are in dependable with the CV analysis. The initial discharge capacity of the battery with BNCNF interlayer is 1182.2 mAh g^{-1} at 0.3 °C, which is 70.71% of the theoretical capacity.

By increasing the current density, the final reversible capacities of the battery with BNCNF interlayer at various current densities of 0.3, 0.5, 1.0 and 3.0 °C are 1030.3, 909.7, 766.3 and 613.3 mAh g^{-1}, respectively. The BNCNF unveils superb absorption for polysulfides underneath the role of the B-S, and N-Li chemical interaction, and especially the formation of $N = B/N$-B structure in the carbon nanofiber framework reinforces the electropositive interaction between B atoms and S_x^{2-} and the electronegative interaction between N atoms and Li^+ cation, simultaneously [68].

Gradually, metal oxides, such as Fe_3O_4, V_2O_5, MnO_2 and TiO_2 [33], and Ti_4O_7 [34], and [35] have displayed the persuasive adsorption with the LiPSs, because of the intense electrostatic attractions between the M–O bond and the LiPSs. Nevertheless, continuing declines of the specific capacity after over the cycling, notably due to the formation of inactive sulfur species inside the interlayer, which impede the ion pathway. Among many metal oxides, V_2O_5 has been used as an interlayer candidate for Li–S application. Very recently, Zhang et al. prepared novel 2D/1D V_2O_5-PP interlayer in which V_2O_5 nanoplates anchored carbon nanofiber interlayer coated on standard polypropylene separator. Figure 5.8 represents the preparation route which composed of electrospinning and hydrothermal method. Also, the reaction mechanism for the formation designed structure clearly demonstrated. SEM images for the CF membrane provided in Fig. 5.8b with a fiber diameter of 100–200 nm. Interestingly, at the high sulfur loading content of about 78.2% applied sample exhibited the low potential of about 0.17, which indicates the excellent performance [69].

5.4.2 Metal Oxides/CNF Composite

Though metal oxides (Except freestanding) have been attracted as cathode due to the adsorption of polysulfide. Then metal oxide nanofibers have been introduced and expected as a versatile route to the large-scale fabrication. Earlier, Tang et al. proposed $Mg_{0.6}Ni_{0.4}O$ hollow nanofibers for the first time. Figure 5.9a reveals SEM images of $Mg_{0.6}Ni_{0.4}O$ nanofibers that hold the length of quite a few hundreds of micrometers the diameter of 100–200 nm. As prepared cathode with $Mg_{0.6}Ni_{0.4}O$ hollow nanofibers calcined at 700 °C exhibited 910 mAh g^{-1} of initial capacity and maintain the enduring capacity of 554 mAh g^{-1} over 20 cycles [70] depicted in Fig. 5.9b. Similarly, TiO_2 nanofibers were prepared by electrospinning method, and TiO_2/S cathode preserves a capacity of 652 mAh/g at 0.1 °C after 200 cycles, corresponding to a capacity retention of 92.7% due to large active sites [71].

Besides, carbon nanofibers incorporated TiO_2, $Li_4Ti_5O_{12}$, and MnO_2 cathode could delay the shuttle effect due to fast Li^+ conductivity and adsorption sites from

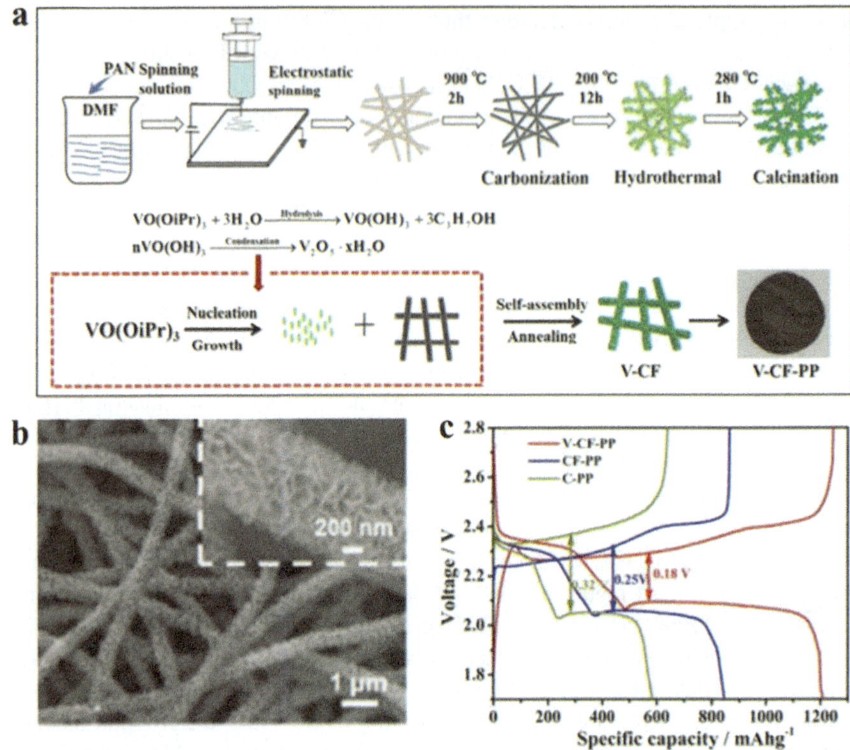

Fig. 5.8 **a** Schematic diagram for the synthesis of 2D/1D V_2O_5 SEM images for the CF membrane the charge/discharge profiles measured at a current rate of 0.2 °C. Reproduced from Ref. [69]. Copyright 2020, by the authors

CNF [73–75]. Meanwhile, $NiCo_2O_4$ nanofibers as carbon-free sulfur immobilizers are hosted to construct sulfur-based composites via electrospinning method. It can be noted, that the SEM and TEM images of $NiCo_2O_4$ nanofibers are presented in Fig. 5.9c with the diameter of 150–200 nm and the wall thickness about 100 and 40 nm. Moreover, they stated that the formation of the hollow structure is ascribed to the decomposition of PAN template during calcination. The S/$NiCo_2O_4$ composite presents a high gravimetric capacity of 1125 mAh g^{-1} see in Fig. 5.9d at 0.1 °C rate, that corresponding to 1500 mAh g^{-1} sulfur based on sulfur mass (high sulfur use of 89.5%), that is 1.2 times that of the S/CNF composite [72]. We also summarized various kind of electrospun based cathode and interlayer materials behaviour for Li–S batteries in Table 5.1

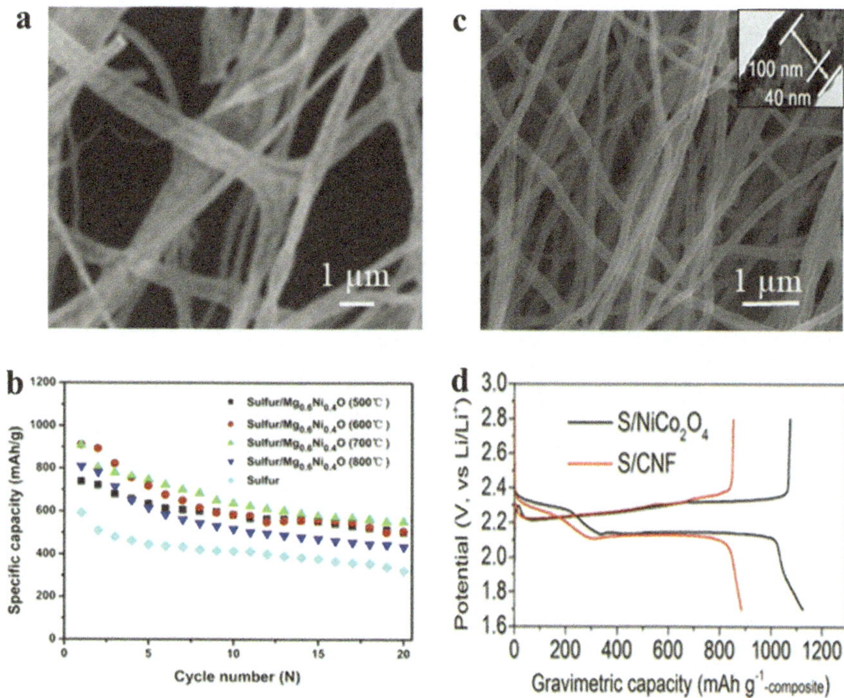

Fig. 5.9 **a** SEM images of Mg$_{0.6}$Ni$_{0.4}$O hollow nanofibers calcined at 700 °C. **b** Cycling performance of Mg$_{0.6}$Ni$_{0.4}$O hollow nanofibers calcined at different temperatures. Reproduced from Ref. [70]. Copyright 2015, Elsevier. **c** SEM images of NiCo$_2$O$_4$ nanofibers and TEM image of NiCo$_2$O$_4$ (inset). **d** Gravimetric capacity plots of S/NiCo$_2$O$_4$ and S/CNF composites at 0.1 °C rate. Reproduced from Ref. [72]. 2019 Wiley–VCH

Table 5.1 Different kind of electrospun cathode and interlayer materials for Li–S batteries

Materials	Precursor	Process	Li–S performance	References
Al$_2$O$_3$ coated FeS$_2$@Carbon fiber	(Fe(C$_5$H$_7$O$_3$)$_3$, /DMF/PAN	1st 250 °C for 2 h in Ar 2nd 600 °C for 5 h in Ar	1110 Wh/kg after 100 cycles	[76]
S/DPAN/KB Nano fiber	DMF/PAN/sulfur/Ketjen Black	1st 300 °C for 1 h in Ar	917 mAh g^{-1} after 150 cycles	[26]

(continued)

Table 5.1 (continued)

Materials	Precursor	Process	Li–S performance	References
S@PCNFs-Cu	$Cu(NO_3)_2 \cdot 3H_2O$/DMF/PAN	1st 280 °C for 3 h in Ar 2nd 400 °C for 1 h in Ar 3rd 800 °C for 6 h in Ar/H_2	680 mA h g^{-1} after 100 cycles at 50 mA/g	[36]
$S_{1-x}Se_x$@PCNFs	DMF/PAN/selenium and sulfur powders	1st 280 °C and 800 °C in N_2	840 mA h g^{-1} after 100 cycles at 0.1 A g^{-1}	[37]
$S_{0.6}Se_{0.4}$@CNFs	DMF/PAN/selenium and sulfur	1st 600 °C for 6 h in Ar	346 mAh g^{-1} after 1000 cycles at 1 A/g	[38]
Flexible Li_2S@NCNF	$Li_2SO_4 \cdot H_2O$/PVP	1st 280 °C for 2 h in air 2nd 800 °C for 2 h in Ar	78% for 200 cycles at 1 °C	[77]
TiO/CNF-S cathodes	Titanium isopropoxide/PVP/ethanol	600 and 950 °C for 6 h in N_2	518 mAh g^{-1} over 200 cycles at 0.5 °C	[42]
NiO/RGO-Sn Hybrid	GO/NiOH/PVP	400 °C in Ar	868 mAh g^{-1} at 150th cycle at 0.1 A^{-1}	[43]
Ti_4O_7/C	TiO_2/DMF/PAN/PVP	1st 280 °C for 2 h in air 2nd 1000 °C for 2 h in N_2	945 mAh g^{-1} after 100 cycles at 0.2 °C	[41]

References

1. P.G. Bruce, Energy storage beyond the horizon: rechargeable lithium batteries. Solid State Ion. **179**(21–26), 752–760 (2008)
2. S.S. Zhang, D. Foster, J. Read, A high energy density lithium/sulfur–oxygen hybrid battery. J. Power Sources **195**(11), 3684–3688 (2010)
3. M. Barghamadi, A. Kapoor, C. Wen, A review on Li-S batteries as a high efficiency rechargeable lithium battery. J. Electrochem. Soc. **160**(8), A1256–A1263 (2013)
4. S. Urbonaite, T. Poux, P. Novák, Progress towards commercially viable Li–S battery cells. Adv. Energy Mater. **5**(16), 1500118 (2015)
5. S. Urbonaite, P. Novák, Importance of 'unimportant' experimental parameters in Li–S battery development. J. Power Sources **249**, 497–502 (2014)
6. X. Ji, K.T. Lee, L.F. Nazar, A highly ordered nanostructured carbon–sulphur cathode for lithium–sulphur batteries. Nat. Mater. **8**(6), 500 (2009)
7. G. Zheng, Y. Yang, J.J. Cha, S.S. Hong, Y. Cui, Hollow carbon nanofiber-encapsulated sulfur cathodes for high specific capacity rechargeable lithium batteries. Nano Lett. **11**(10), 4462–4467 (2011)
8. Y.V. Mikhaylik, J.R. Akridge, Polysulfide shuttle study in the Li/S battery system. J. Electrochem. Soc. **151**(11), A1969–A1976 (2004)

9. S.-E. Cheon, K.-S. Ko, J.-H. Cho, S.-W. Kim, E.-Y. Chin, H.-T. Kim, Rechargeable lithium sulfur battery I. Structural change of sulfur cathode during discharge and charge. J. Electrochem. Soc. **150**(6), A796–A799 (2003)

10. F. Wu, L. Shi, D. Mu, H. Xu, B. Wu, A hierarchical carbon fiber/sulfur composite as cathode material for Li–S batteries. Carbon **86**, 146–155 (2015)

11. J.H. Yun, J.-H. Kim, D.K. Kim, H.-W. Lee, Suppressing polysulfide dissolution via cohesive forces by interwoven carbon nanofibers for high-areal-capacity lithium–sulfur batteries. Nano Lett. **18**(1), 475–481 (2017)

12. Y. Yang, M.T. McDowell, A. Jackson, J.J. Cha, S.S. Hong, Y. Cui, New nanostructured Li2S/silicon rechargeable battery with high specific energy. Nano Lett. **10**(4), 1486–1491 (2010)

13. X. He, J. Ren, L. Wang, W. Pu, C. Wan, C. Jiang, Li/S lithium ion cell using graphite as anode. ECS Trans. **2**(8), 47–49 (2007)

14. J. Hassoun, B. Scrosati, A high-performance polymer tin sulfur lithium ion battery. Angew. Chem. Int. Ed. **49**(13), 2371–2374 (2010)

15. Y.S. Nimon, M.-Y. Chu, S.J. Visco, Coated lithium electrodes, Google Patents, 2003

16. S. Liu, J. Yang, L. Yin, Z. Li, J. Wang, Y. Nuli, Lithium-rich Li_2. 6BMg 0. 05 alloy as an alternative anode to metallic lithium for rechargeable lithium batteries. Electrochim. Acta **56**(24), 8900–8905 (2011)

17. R.A. Guidotti, P.J. Masset, Thermally activated ("thermal") battery technology: Part IV. Anode materials. J. Power Sources **183**(1), 388–398 (2008)

18. P. Zeng, Y. Han, X. Duan, G. Jia, L. Huang, Y. Chen, A stable graphite electrode in super-concentrated LiTFSI-DME/DOL electrolyte and its application in lithium-sulfur full battery. Mater. Res. Bull. **95**, 61–70 (2017)

19. M. Li, X. Liu, Q. Li, Z. Jin, W. Wang, A. Wang, Y. Huang, Y. Yang, P4S10 modified lithium anode for enhanced performance of lithium–sulfur batteries. J. Energy Chem. **41**, 27–33 (2020)

20. H. Wu, S.A. Shevlin, Q. Meng, W. Guo, Y. Meng, K. Lu, Z. Wei, Z. Guo, Flexible and binder-free organic cathode for high-performance lithium-ion batteries. Adv. Mater. **26**(20), 3338–3343 (2014)

21. Y. Yang, G. Zheng, Y. Cui, Nanostructured sulfur cathodes. Chem. Soc. Rev. **42**(7), 3018–3032 (2013)

22. M. Zhang, Q. Meng, A. Ahmad, L. Mao, W. Yan, Z. Wei, Poly (3, 4-ethylenedioxythiophene)-coated sulfur for flexible and binder-free cathodes of lithium–sulfur batteries. J. Mater. Chem. A **5**(33), 17647–17652 (2017)

23. A. Manthiram, Y. Fu, S.-H. Chung, C. Zu, Y.-S. Su, Rechargeable lithium–sulfur batteries. Chem. Rev. **114**(23), 11751–11787 (2014)

24. C. Zu, Y. Fu, A. Manthiram, Highly reversible Li/dissolved polysulfide batteries with binder-free carbon nanofiber electrodes. J. Mater. Chem. A **1**(35), 10362–10367 (2013)

25. S. Lu, Y. Chen, X. Wu, Z. Wang, L. Lv, W. Qin, L. Jiang, Binder-free cathodes based on sulfur–carbon nanofibers composites for lithium–sulfur batteries. RSC Adv. **4**(35), 18052–18054 (2014)

26. S. Kalybekkyzy, A. Mentbayeva, M.V. Kahraman, Y. Zhang, Z. Bakenov, Flexible S/DPAN/KB nanofiber composite as binder-free cathodes for Li-S batteries. J. Electrochem. Soc. **166**(3), A5396–A5402 (2019)

27. W. Li, Q. Zhang, G. Zheng, Z.W. Seh, H. Yao, Y. Cui, Understanding the role of different conductive polymers in improving the nanostructured sulfur cathode performance. Nano Lett. **13**(11), 5534–5540 (2013)

28. Z. Li, G. Ma, R. Ge, F. Qin, X. Dong, W. Meng, T. Liu, J. Tong, F. Jiang, Y. Zhou, Free-standing conducting polymer films for high-performance energy devices. Angew. Chem. Int. Ed. **55**(3), 979–982 (2016)

29. Y. Zhu, X. Fan, L. Suo, C. Luo, T. Gao, C. Wang, Electrospun FeS2@ carbon fiber electrode as a high energy density cathode for rechargeable lithium batteries. ACS Nano **10**(1), 1529–1538 (2015)

30. R. Fong, J. Dahn, C. Jones, Electrochemistry of pyrite-based cathodes for ambient temperature lithium batteries. J. Electrochem. Soc. **136**(11), 3206–3210 (1989)
31. C. Jones, P. Kovacs, R. Sharma, R. McMillan, Iron-57 Moessbauer spectroscopy of reduced cathodes in the lithium/iron disulfide battery system: evidence for superparamagnetism. J. Phys. Chem. **94**(2), 832–836 (1990)
32. C. Jones, P. Kovacs, R. Sharma, R. McMillan, An iron-57 Moessbauer study of the intermediates formed in the reduction of iron disulfide in the lithium/iron disulfide battery system. J. Phys. Chem. **95**(2), 774–779 (1991)
33. S.B. Son, T.A. Yersak, D.M. Piper, S.C. Kim, C.S. Kang, J.S. Cho, S.S. Suh, Y.U. Kim, K.H. Oh, S.H. Lee, A stabilized PAN-FeS2 cathode with an EC/DEC liquid electrolyte. Adv Energy Mater **4**(3), 1300961 (2014)
34. C. Ye, L. Zhang, C. Guo, D. Li, A. Vasileff, H. Wang, S.Z. Qiao, A 3D hybrid of chemically coupled nickel sulfide and hollow carbon spheres for high performance lithium–sulfur batteries. Adv. Func. Mater. **27**(33), 1702524 (2017)
35. S. Zheng, F. Yi, Z. Li, Y. Zhu, Y. Xu, C. Luo, J. Yang, C. Wang, Copper-stabilized sulfur-microporous carbon cathodes for Li–S batteries. Adv. Func. Mater. **24**(26), 4156–4163 (2014)
36. L. Zeng, Y. Jiang, J. Xu, M. Wang, W. Li, Y. Yu, Flexible copper-stabilized sulfur–carbon nanofibers with excellent electrochemical performance for Li–S batteries. Nanoscale **7**(25), 10940–10949 (2015)
37. L. Zeng, Y. Yao, J. Shi, Y. Jiang, W. Li, L. Gu, Y. Yu, A flexible S1− xSex@ porous carbon nanofibers (x≤ 0.1) thin film with high performance for Li-S batteries and room-temperature Na-S batteries. Energy Storage Mater. **5**, 50–57 (2016)
38. Y. Yao, L. Zeng, S. Hu, Y. Jiang, B. Yuan, Y. Yu, Binding S06Se0. 4 in 1D carbon nanofiber with C S bonding for high-performance flexible Li–S batteries and Na–S batteries. Small 13(19), 1603513 (2017)
39. R. Zhuang, S. Yao, X. Shen, T. Li, A freestanding MoO2-decorated carbon nanofibers interlayer for rechargeable lithium sulfur battery. Int. J. Energy Res. **43**(3), 1111–1120 (2019)
40. M. Liu, J. Hou, J. Xiang, X. Shen, K. Luan, Y. Zhang, Effect of non-woven Al2O3/C nanofibers as functional interlayer on electrochemical performance of lithium-sulfur batteries. J. Nanosci. Nanotechnol. **18**(11), 7824–7829 (2018)
41. Y. Guo, J. Li, R. Pitcheri, J. Zhu, P. Wen, Y. Qiu, Electrospun Ti4O7/C conductive nanofibers as interlayer for lithium-sulfur batteries with ultra long cycle life and high-rate capability. Chem. Eng. J. **355**, 390–398 (2019)
42. A. Singh, V. Kalra, TiO phase stabilized into freestanding nanofibers as strong polysulfide immobilizer in Li–S batteries: evidence for lewis acid-base interactions. ACS Appl. Mater. Interfaces **10**(44), 37937–37947 (2018)
43. C. Li, S. Dong, D. Guo, Z. Zhang, M. Wang, L. Yin, Ternary NiO/RGO-Sn hybrid flexible free-standing film as interlayer for lithium-sulfur batteries with improved performance. Electrochim. Acta **251**, 43–50 (2017)
44. R. Singhal, S.-H. Chung, A. Manthiram, V. Kalra, A free-standing carbon nanofiber interlayer for high-performance lithium–sulfur batteries. J. Mater. Chem. A **3**(8), 4530–4538 (2015)
45. Y. Chen, X. Li, K.-S. Park, J. Hong, J. Song, L. Zhou, Y.-W. Mai, H. Huang, J.B. Goodenough, Sulfur encapsulated in porous hollow CNTs@ CNFs for high-performance lithium–sulfur batteries. J. Mater. Chem. A **2**(26), 10126–10130 (2014)
46. S. Han, X. Pu, X. Li, M. Liu, M. Li, N. Feng, S. Dou, W. Hu, High areal capacity of Li-S batteries enabled by freestanding CNF/rGO electrode with high loading of lithium polysulfide. Electrochim. Acta **241**, 406–413 (2017)
47. D. Yang, W. Ni, J. Cheng, Z. Wang, T. Wang, Q. Guan, Y. Zhang, H. Wu, X. Li, B. Wang, Flexible three-dimensional electrodes of hollow carbon bead strings as graded sulfur reservoirs and the synergistic mechanism for lithium–sulfur batteries. Appl. Surf. Sci. **413**, 209–218 (2017)
48. J. Wang, G. Yang, J. Chen, Y. Liu, Y. Wang, C.Y. Lao, K. Xi, D. Yang, C.J. Harris, W. Yan, Flexible and high-loading lithium–sulfur batteries enabled by integrated three-in-one fibrous membranes. Adv. Energy Mater. (2019)

49. J. He, G. Hartmann, M. Lee, G.S. Hwang, Y. Chen, A. Manthiram, Freestanding 1T MoS 2/graphene heterostructures as a highly efficient electrocatalyst for lithium polysulfides in Li–S batteries. Energy Environ. Sci. **12**(1), 344–350 (2019)

50. Q. Hao, G. Cui, Y. Zhang, J. Li, Z. Zhang, Novel MoSe2/MoO2 heterostructure as an effective sulfur host for high-performance lithium/sulfur batteries. Chem. Eng. J. **381**, 122672 (2020)

51. R. Zhuang, S. Yao, M. Liu, J. Wu, X. Shen, T. Li, β-molybdenum carbide/carbon nanofibers as a shuttle inhibitor for lithium-sulfur battery with high sulfur loading. Int. J. Energy Res. (2019)

52. R. Zhuang, S. Yao, X. Shen, T. Li, S. Qin, J. Yang, Electrospun β-Mo 2 C/CNFs as an efficient sulfur host for rechargeable lithium sulfur battery. J. Mater. Sci.: Mater. Electron. **30**(5), 4626–4633 (2019)

53. Z. Sun, J. Zhang, L. Yin, G. Hu, R. Fang, H.-M. Cheng, F. Li, Conductive porous vanadium nitride/graphene composite as chemical anchor of polysulfides for lithium-sulfur batteries. Nat. Commun. **8**, 14627 (2017)

54. X. Zhao, M. Liu, Y. Chen, B. Hou, N. Zhang, B. Chen, N. Yang, K. Chen, J. Li, L. An, Fabrication of layered Ti 3 C 2 with an accordion-like structure as a potential cathode material for high performance lithium–sulfur batteries. J. Mater. Chem. A **3**(15), 7870–7876 (2015)

55. J.-Q. Huang, B. Zhang, Z.-L. Xu, S. Abouali, M.A. Garakani, J. Huang, J.-K. Kim, Novel interlayer made from Fe3C/carbon nanofiber webs for high performance lithium–sulfur batteries. J. Power Sources **285**, 43–50 (2015)

56. J.H. Kim, J. Seo, J. Choi, D. Shin, M. Carter, Y. Jeon, C. Wang, L. Hu, U. Paik, Synergistic ultrathin functional polymer-coated carbon nanotube interlayer for high performance lithium–sulfur batteries. ACS Appl. Mater. Interfaces **8**(31), 20092–20099 (2016)

57. S.S. Zhang, D.T. Tran, Z. Zhang, Poly (acrylic acid) gel as a polysulphide blocking layer for high-performance lithium/sulphur battery. J. Mater. Chem. A **2**(43), 18288–18292 (2014)

58. M. Liu, N. Deng, J. Ju, L. Fan, L. Wang, Z. Li, H. Zhao, G. Yang, W. Kang, J. Yan, A review: electrospun nanofiber materials for lithium-sulfur batteries. Adv. Funct. Mater. (2019)

59. H. Wei, J. Ma, B. Li, Y. Zuo, D. Xia, Enhanced cycle performance of lithium–sulfur batteries using a separator modified with a PVDF-C layer. ACS Appl. Mater. Interfaces **6**(22), 20276–20281 (2014)

60. Y.-S. Su, A. Manthiram, Lithium–sulphur batteries with a microporous carbon paper as a bifunctional interlayer. Nat. Commun. **3**, 1166 (2012)

61. Y.-S. Su, A. Manthiram, A new approach to improve cycle performance of rechargeable lithium–sulfur batteries by inserting a free-standing MWCNT interlayer. Chem. Commun. **48**(70), 8817–8819 (2012)

62. S.H. Chung, A. Manthiram, A natural carbonized leaf as polysulfide diffusion inhibitor for high-performance lithium-sulfur battery cells. Chemsuschem **7**(6), 1655–1661 (2014)

63. X. Wang, Z. Wang, L. Chen, Reduced graphene oxide film as a shuttle-inhibiting interlayer in a lithium–sulfur battery. J. Power Sources **242**, 65–69 (2013)

64. S.H. Chung, P. Han, R. Singhal, V. Kalra, A. Manthiram, Electrochemically stable rechargeable lithium–sulfur batteries with a microporous carbon nanofiber filter for polysulfide. Adv. Energy Mater. **5**(18), 1500738 (2015)

65. J. Wang, Y. Yang, F. Kang, Porous carbon nanofiber paper as an effective interlayer for high-performance lithium-sulfur batteries. Electrochim. Acta **168**, 271–276 (2015)

66. Y. Peng, Y. Zhang, Y. Wang, X. Shen, F. Wang, H. Li, B.-J. Hwang, J. Zhao, Directly coating a multifunctional interlayer on the cathode via electrospinning for advanced lithium–sulfur batteries. ACS Appl. Mater. Interfaces **9**(35), 29804–29811 (2017)

67. T. Gao, T. Le, Y. Yang, Z. Yu, Z. Huang, F. Kang, Effects of electrospun carbon nanofibers' interlayers on high-performance lithium-sulfur batteries. Materials **10**(4), 376 (2017)

68. J. Zhu, R. Pitcheri, T. Kang, C. Jiao, Y. Guo, J. Li, Y. Qiu, A polysulfide-trapping interlayer constructed by boron and nitrogen co-doped carbon nanofibers for long-life lithium sulfur batteries. J. Electroanal. Chem. **833**, 151–159 (2019)

69. Z. Zhang, G. Wu, H. Ji, D. Chen, D. Xia, K. Gao, J. Xu, B. Mao, S. Yi, L. Zhang, 2D/1D V2O5 nanoplates anchored carbon nanofibers as efficient separator interlayer for highly stable lithium-sulfur battery. Nanomaterials **10**(4), 705 (2020)

70. H. Tang, S. Yao, M. Jing, X. Wu, J. Hou, X. Qian, D. Rao, X. Shen, X. Xi, K. Xiao, Mg0. 6Ni0. 4O hollow nanofibers prepared by electrospinning as additive for improving electrochemical performance of lithium–sulfur batteries. J. Alloys Compd. **650**, 351–356 (2015)
71. X. Shan, Z. Guo, X. Zhang, J. Yang, L. Duan, Mesoporous TiO 2 nanofiber as highly efficient sulfur host for advanced lithium-sulfur batteries. Chin. J. Mech. Eng. **32**(1), 60 (2019)
72. Y.T. Liu, D.D. Han, L. Wang, G.R. Li, S. Liu, X.P. Gao, NiCo2O4 nanofibers as carbon-free sulfur immobilizer to fabricate sulfur-based composite with high volumetric capacity for lithium-sulfur battery. Adv. Energy Mater. **9**(11), 1803477 (2019)
73. S. Yao, H. Tang, M. Liu, L. Chen, M. Jing, X. Shen, T. Li, J. Tan, TiO2 nanoparticles incorporation in carbon nanofiber as a multi-functional interlayer toward ultralong cycle-life lithium-sulfur batteries. J. Alloys Compd. **788**, 639–648 (2019)
74. D. An, L. Shen, D. Lei, L. Wang, H. Ye, B. Li, F. Kang, Y.-B. He, An ultrathin and continuous Li4Ti5O12 coated carbon nanofiber interlayer for high rate lithium sulfur battery. J. Energy Chem. **31**, 19–26 (2019)
75. Z. Liu, B. Liu, P. Guo, X. Shang, M. Lv, D. Liu, D. He, Enhanced electrochemical kinetics in lithium-sulfur batteries by using carbon nanofibers/manganese dioxide composite as a bifunctional coating on sulfur cathode. Electrochim. Acta **269**, 180–187 (2018)
76. Y. Zhu, X. Fan, L. Suo, C. Luo, T. Gao, C. Wang, Electrospun FeS2@ carbon fiber electrode as a high energy density cathode for rechargeable lithium batteries. ACS Nano **10**(1), 1529–1538 (2016)
77. M. Yu, Z. Wang, Y. Wang, Y. Dong, J. Qiu, Freestanding flexible Li2S paper electrode with high mass and capacity loading for high-energy Li–S batteries. Adv. Energy Mater. **7**(17), 1700018 (2017)

Chapter 6
Electrospinning of Nanofibers for Zn-Air Battery

Abstract Zinc-air battery has been attracted technology among all metal-air system for commercial use because of larger storage capacity at a fraction of the cost compared to LIBs. Zinc-air battery signifies one of the most viable future options to powering electric vehicles. But some technical difficulties connected with them have yet to be resolved. In this chapter, we first discussed the fundamentals and working principle, and then we overviewed challenges and recent advancement related to zinc-air battery studies. Secondly, we summarized perspectives on the contemporary electrode materials for both anode and cathode. Our mainstream is the review of electrospun nanofibers contribution towards the high performance of the zinc-air battery. Meanwhile, the OER and ORR electrocatalyst activities, preparation techniques and their power densities have been compared and summarized.

Keywords Zinc-air battery · Fiber electrodes · Porous structure · Electrocatalysts

6.1 Zn-Air Battery Working Principle and Cell Structure

Zn-Air batteries have powerful potential for use as an alternative energy storage system due to many of advantages like plentiful, cheap and low equilibrium potential. It is included of a negative zinc electrode, a membrane separator and a positive air electrode assembled in an alkaline electrolyte as illustrated with a schematic framework of a Zn-Air battery in Fig. 6.1.

During the battery discharge, the oxidation of zinc happens, giving rise to soluble zincate ions (i.e. $Zn(OH)_4^{2-}$). This method typically continues until they are supersaturated in the electrolyte, after which the zincate ions decompose to insoluble zinc oxide [1, 2]. The following reaction describes the working principle of Zn-air battery.

At anode:

$$Zn + 4OH^- \rightarrow Zn(OH)_4^{2-} + 2e^- \tag{6.1}$$

$$Zn(OH)_4^{2-} \rightarrow ZnO + H_2O + 2OH^-$$

S. Peng and P. R. Ilango, *Electrospinning of Nanofibers for Battery Applications*,
https://doi.org/10.1007/978-981-15-1428-9_6

Fig. 6.1 A schematic outline of a Zn-air battery. Reproduced from Ref. [1]. Copyright 2014, The Royal Society of Chemistry

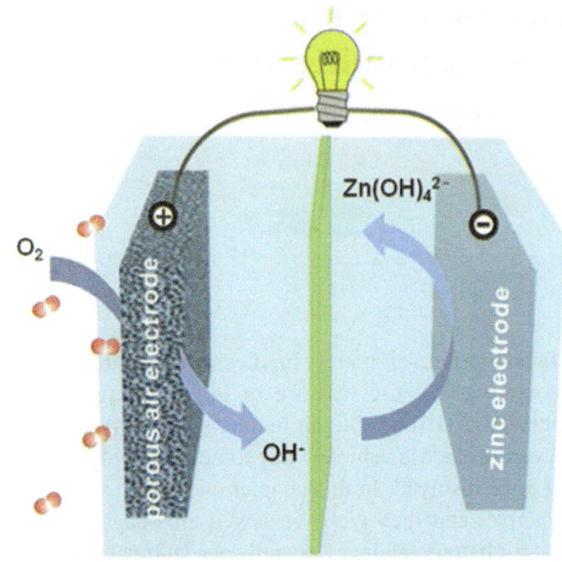

At cathode:

$$O_2 + 4e^- + 2H_2O \rightarrow 4OH^-$$ (6.2)

Overall reaction:

$$2Zn + 2H_2O \rightarrow 2ZnO$$ (6.3)

Parasitic reaction:

$$Zn + 2H_2O \rightarrow Zn(OH)_2 + H_2$$ (6.4)

Corresponding with the oxidation reaction at the negative electrode, an undesired parasitic reaction between zinc and water can occur ensuing in hydrogen gas generation. It causes a gradual self-corrosion of the zinc metal and lowers the active material utilization. At the positive electrode, oxygen from the adjacent atmosphere infuses the porous gas diffusion electrode (GDE). It gets reduced on the surface of the electrocatalyst particles in close contact with the electrolyte. The oxygen reduction reaction (ORR) taking place at the air electrode of zinc-air batteries is similar to the ORR in alkaline hydrogen fuel cells with hydroxide ions being the first product [1].

6.2 Perspectives on Material Development

6.2.1 Anode

Metallic zinc (Zn) has been the anode material of choice for many primary systems such as zinc-carbon, zinc–manganese dioxide, zinc-nickel and zinc-air since the discovery of the first battery by Volta in 1796 [1]. It retains an exceptional usual of features comprising low equivalent weight, reversibility, high specific energy density, abundance and low toxicity. It is the most electropositive metal that is constant in aqueous and alkaline media short of substantial corrosion. In general, during the discharge process oxidation of Zn occurs. Even though the Zn corrosion rate is slower than aluminium in alkaline solution. The corrosion process is the main reason produce hydrogen gas (Eq. 6.4). Hence it is more accountable to look at the electrochemical characteristic of Zn metal in alkaline solution [3]. Over the past decades, noteworthy effort to develop large systems owing to their high-power requirement, the anodes in these systems typically use dendritic forms of zinc powder. So far, many Zn metal-based anodes have been prepared like zinc metal with nickel and alloy with other metals like Hg, Pb and Cd. Remarkably fibrous materials were recommended long ago as having advantages over powders. Teck Cominco has established a new way to make zinc anodes using fibrous materials for high power applications of alkaline batteries and large size zinc–air cells and their results shows improved performance. Figure 6.2 describes the Zn nanofibers which were fabricated by a spin cast method.

Figure 6.2b shows a discharging profile at 1 A continuous current for a C cell assembled with a fibrous anode in contrast to that for an original commercial C cell. The cell with the fibrous anode implemented greatly than the commercial cell. Captivating the capacity at 0.8 V as a measurement, the cell with the fibrous anode had about 30% more capacity than the commercial cell. The test cell displayed higher voltage than the commercial cell, and the variance amplified with time, which represents that the potential loss is less in the fibrous anode, mainly near the finish

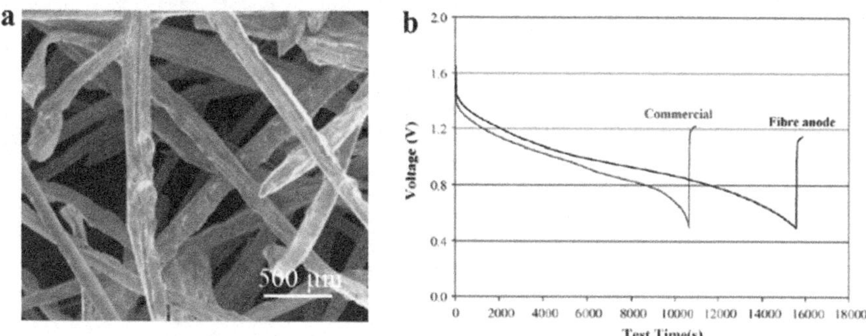

Fig. 6.2 **a** SEM images of a typical Zn fibers. **b** Discharge curves for an original C cell and a cell using Zn fiber anode. Reproduced from Ref. [4] Copyright 2006, Elsevier

of discharge [4]. Later, while preparing a Zn anode with powder, mercury has been added to the zinc anode to give better electrical conductivity between the zinc particles and the current collectors.

6.2.2 Cathode

Cathode in Zn-air battery classified into two parts (i) gas diffusion layer and catalytic layer see Fig. 6.3. Gas diffusion layer stands on the backside and looks to the open air while the catalytic active layer shields on the surface of the current collector and contacts with electrolyte [1]. The concept of using oxygen in a zinc-air battery requires the air electrode to have both proper catalysts for oxygen reduction reaction and a highly porous structure [3]. Hence porous structure is advantageous for air electrode. Catalysts for the oxygen evolution and oxygen reduction reactions (OER/ORR) are at the central of the leading Zn-air batteries.

6.3   Binder-Free Electrodes

In recent times, the binder-free electrode has shown many essential advantages, such as controlled micro/nanostructures, shortened diffusion paths, and high conductivities, then powder samples in potential applications. Indeed, this type of electrodes performance is quite outstanding properties, as mentioned above [5]. As of many procedures have been introduced to prepare flexible electrodes. For example, Lee et al. demonstrated cutting-edge air electrode with functionality and practicality for

Fig. 6.3 The structure of gas diffusion electrode. Reproduced from Ref. [1]. Copyright 2014, The Royal Society of Chemistry

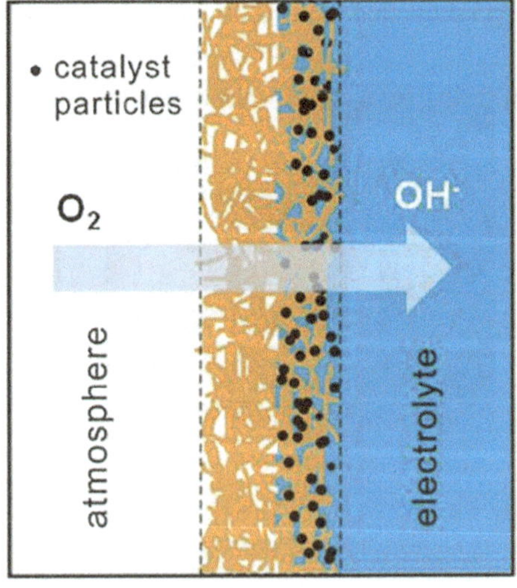

long term rechargeable zinc-air battery applications through 3D binder-free electrodes. The scheme provided in Fig. 6.4a and demonstrating extended cycling of 600 h with charge and discharge potential retentions of 97% and 94%, respectively. More significant discharge and charge potentials of the innovative SS mesh electrode are apparent in the galvanodynamic discharge and charge polarization profiles beyond 20 mA cm^{-2} [6] shown in Fig. 6.4b. Later, Chen et al. proposed ultrathin mesoporous Co_3O_4 layers on the surface of carbon fibers and tested for the zinc-air battery test. As prepared binder-free electrode displays outstanding rechargeable performance ($\approx$1.03 V discharge voltage and $\approx$1.95 V charge voltage at 2 mA cm^{-2}), with a high energy density of 546 W h kg^{-1}. Matched to the commercial Co_3O_4/CC electrode, the flexible Zn–air battery using ultrathin Co_3O_4/CC electrode reveals extremely high mechanical stability. It can function without performance loss, even under serious and frequent deformation [7]. Interestingly, Co_3O_4 decorated CNF as binder-free electrode reported with low charge–discharge voltage cap with higher stability and their SEM images demonstrated in Fig. 6.4c following cycling performance indicates its good efficient catalytic behaviour [8] shown in Fig. 6.4d. A very recent report says that N-doped porous carbon flakes grown on nanofibers displays the dual characteristics of outstanding catalytic activity (reversible oxygen overpotential of 0.75 V) and high stability, due to the significantly enhanced active sites [9].

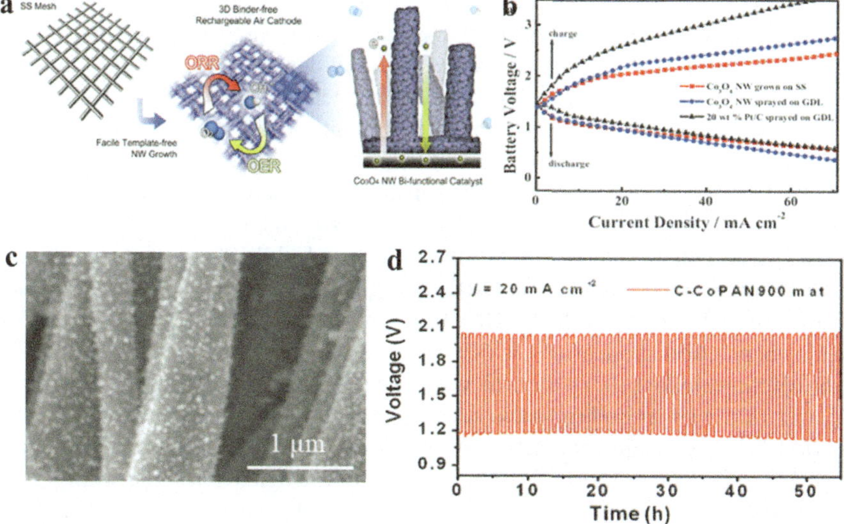

Fig. 6.4 **a** Schematic illustration of the growth of 3D rechargeable Co_3O_4 NW air cathode for bifunctional catalysis of ORR and OER. **b** Galvanodynamic discharge and charge polarization curves obtained using air under ambient conditions of Co_3O_4 NWs grown on SS mesh. Reproduced from Ref. [6]. Copyright 2013, Wiley–VCH. **c** Representative SEM image of C-CoPAN900. **d** Cycling performances of the ZABs at a high current density of 20 mA cm^{-2}. Reproduced from Ref. [8]. Copyright 2015, The Royal Society of Chemistry

6.4 Free-Standing Electrodes

Usually, the building approach for such binder-free structures can be classified into two types, such as template-free fabrication, during which catalytic materials or their precursors are combined with organic sources to form a free-standing structure. On the other hand, organic free-standing structure can be used as a substrate for supporting functional materials or their precursors. Indeed, high temperature calcination treatment is frequently essential in both methods to get electroactive free-standing electrodes. For instance, electrospinning is a famous process in preparing free-standing nanofibrous structures [10]. Also, the free-standing electrode is necessarily required as air cathode to show flexible and overcome the ORR/OER essential challenges during Zn-air battery applications [11]. For example, transition metal carbides are the strong applicants for extensive use in hydrogen evolution reactions and oxygen reduction reactions (HER and ORR), showing a catalytic activity similar to Pt-based electrochemical catalysts. To support this, Zhao et al. prepared Fe_3C@NCNTs-NCNFs in which FeC_3 nanoparticles entrapped at the edge of nitrogen-enriched carbon nanotubes aligned on CNFs as free-standing electrode possess the overpotential of 284 mV and a low tafel slope of 56 mV dec^{-1} [12]. Similarly, 3D CNT/CNFs based free-standing long-term steadiness and tolerance to methanol crossover for the ORR in an alkaline medium [13]. Metal–organic framework (MOF) derived carbon nanofibers also has a large notice in the electrocatalysis. Interestingly, Zn, Co and Fe based MOF derived materials have great impact on ORR catalysts [14, 15]. Not long ago, Fe-NHCFs based free-standing developed by Wu et al. and reported a discharge current density of 90 mA cm^{-2} and an output peak power density of 61 mW cm^{-2} when attempting with zinc-air battery. As prepared electrode TEM images (see Fig. 6.5a) confirms the nanofibers constructed with numerous hollow nanoparticles while Fe-NHCFs reduced to 230 mm after calcination which is confirmed by SEM. The galvanostatic discharge tests for Fe-NHCFs were carried out at various current densities (5, 10, and 15 mA cm^{-2}) shown in Fig. 6.5b. It is noted that no voltage drop was detected while process, which inferred a satisfactory catalytic stability for ORR [15] and all the gained results specified that the exceptional stability of the Fe-NHCFs electrode can thus allow long-term work loading in practical applications, which was represented by unremitting lighting of a light-emitting diode (LED) pattern (3 V) by two integrated zinc–air batteries in series for over 12 h without no brightness decay (see in Fig. 6.5c). Newly, PAN@ZIF-67 hybrid fabricated in-situ approach with electro spun nanofibers. Figure 6.5d depicts FESEM images of C-PAN@ZIF-67, is a randomly entangled fibrous network which highly desirable as flexible electrodes of devices in the wearables. Accordingly, a 28% higher peak power density (63 vs. 49 mW cm^{-2}) was achieved by ZnAB with C-PAN@ZIF-67 over that of its counterpart using Pt/C see in Fig. 6.5e. More importantly, Fig. 6.5f proving the satisfactory bendability of the cell. This study suggested that as prepared free-standing electrode has good conductivity and flexibility, which further ORR activity enhancement by synergistic couplings between the two [16].

6.5 Types of Catalysts

6.5.1 ORR Electrocatalysts

The oxygen reduction reaction (ORR) at the cathode is one of the vital factors impacting the performance of fuel cells and metal-air batteries [17]. Typically, plantinum (Pt)-based materials are extensively used as the electrocatalysts for ORR. Alternatively, Pt-free catalysis transition metal chalcogenides, Pd, Ag Co and porous carbon and activated carbon-based materials have been found to have electrocatalytic activity on ORR [18–21]. In recent years, excessive steps have been dedicated to the synthesis of 1D carbon materials with hybrid nanostructures in several research areas, such as catalysis, electrode and sensing. Essentially, electrospinning is a facile and cost-effective technique to formulate identical fibers, which hold large length-diameter ratio, high specific surface area, and excellent flexibility [22]. Evidently, Tungsten carbide (WC) nanofibers with a controlled diameter of 40–300 nm were fabricated by electrospinning method, which showed remarkable ORR activities [23]. Earlier, Park et al. reported porous with nitrogen-doped carbon nanofiber as ORR catalysts and tested for Zn-air battery which exhibits a power density of 194 $mWcm^{-2}$.It is stated that the new formation of inner pores and outer etched surface [20]. Likewise, Zhao et al. proposed HP-Fe–N/CNFs interconnected hierarchically porous carbon nanofibers simultaneously doped with nitrogen and iron, which has a high surface area of 569.6 m^2/g. The schematic preparation method has been demonstrated in Fig. 6.6a. Profiting from the enhanced mass transfer and exploitation of active sites endorsed to interconnected hierarchical porous structures, HP-Fe-N/CNFs reveal admirable ORR catalytic activity in alkaline media, with a as good as onset potential and half-wave potential but superior selectivity, stability, and tolerance in contradiction of methanol to commercial 30 wt% Pt/C. Principally, when applied in an assembled Zn-air battery, HP-Fe-N/CNFs outperform 30 wt% Pt/C in power density and long-term stability, explicitly (see in Fig. 6.6b) showing their capable real-world application [24]. Besides, Zeolitic imidazolate frameworks (ZIFs) derived composite nanofibers Co/CoOx-N-C) [25], PMSEPE/MIL-101(Fe) [26] and Zn/Co-ZIFs/PAN [27] have been proposed as efficient ORR catalysts for Zn-Air batteries. Undoubtedly, they have shown high power density and remarkable durability. Electrospun nanofibers are also performing excellently when it's applied into flexible Zn-air battery due to their intrinsic topography and high surface area [28]. Particular interest in flexible Zn-air battery, the Zn/Co–N@ PCNFs superior ORR catalytic activity. As prepared Zn/Co–N@ PCNFs-800 ORR catalysts SEM images illustrated in Fig. 6.6c with folded surface after calcination to that of commercial 20 wt% Pt/C, in terms of its onset potential (0.98 V vs. RHE), half-wave potential (0.89 V vs. RHE), and limiting current density (5.26 mA cm^{-2}) [27].

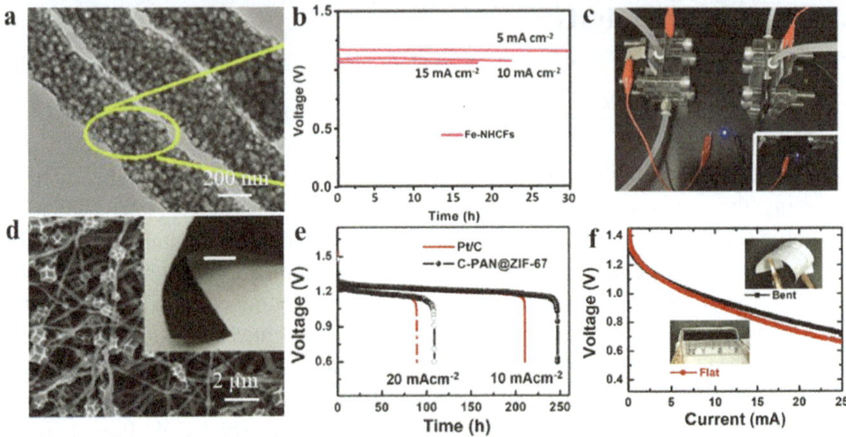

Fig. 6.5 **a** TEM images of Fe-NHCFs. **b** Galvanostatic discharge curves of Zn–air batteries using Fe-NHCFs as ORR catalysts and the KOH electrolyte at various current densities. **c** Photographs of lighting of an LED pattern. Reproduced from Ref. [15]. Copyright 2019, Wiley–VCH. **d** FESEM image of C-PAN@ZIF-67. **e** Galvanostatic discharge curves of ZnABs with C-PAN@ZIF-67. **f** Polarization curves of the flexible Zn-air cell at flat and bent states. Reproduced from Ref. [16]. Copyright 2019, Elsevier

6.5.2 OER Electrocatalysts

The OER is an applied process in many energy storage applications such as water electrolysis, co-electrolysis, and metal-air batteries. Miserably, the OER reaction is marked by slow reaction kinetics and large overpotentials [29]. The justification of the slow kinetics of OER is that the reaction comprises of multiple elementary reaction steps linking four-electron transfer and producing many middle species. As a result of its complexity, several different OER mechanisms have been advised. The following principle is the most recognized one on metal oxide surface in acidic media.

$$S + H_2O \rightarrow S - OH_{ads} + H^+ + e^- \tag{6.5}$$

$$S - OH_{ads} \rightarrow S - O_{ads} + H^+ + e^- \tag{6.6}$$

$$S - O_{ads} + H_2O \rightarrow S - OOH_{ads} + H^+ + e^- \tag{6.7}$$

$$S - OOH_{ads} \rightarrow S + O_2 + H^+ + e^- \tag{6.8}$$

where S is the surface active site of a heterogeneous catalyst, and $-OH_{ads}$, $-O_{ads}$, and $-OOH_{ads}$ are hydroxyl, oxygen, and hydroperoxyl species adsorbed on the surface active site, respectively [30]. Over the past years, so many research have been carried out with the purpose of classifying appropriate catalysts to upgrade the reaction rate and boost the stability for oxygen evolution [29]. IrO_2 and RuO_2 are known as two of the most capable OER catalysts bye the reason for their low overpotential values during the reaction [31]. The electrospun nanofibers based electrocatalysts have been reported such as NiO [31], $La_{0.5}Sr_{0.5}Co_{0.8}Fe_{0.2}O_3$, Fe_3C [12], Cu/PVdF-HFP [32] and $CaMnO_3$ [33] respectively. Regard to zinc-air battery, the above-mentioned catalysts outperformed well in terms of durability and current density. Among them, Zhao et al. proposed Fe_3C@NCNTs-NCNFs in which Fe_3C nanoparticles encapsulated at the tip of nitrogen enriched carbon nanotubes which are aligned on one-dimensional nitrogen-doped carbon nanofibers by a scalable electrospinning technique, as a high performance OER catalyst. The scheme of the fabrication of Fe_3C@NCNTs-NCNFs as shown in Fig. 6.7a. The SEM image of Fe_3C@NCNTs-NCNFs shows the continuous and interconnected nanofibers forming a hierarchical 3D architecture. The average diameter of the electrospun fibers is around 500 nm while Fe_3C@NCNTs arrays has 150–200 nm see in Fig. 6.7b. The pure NCNFs exhibit very poor OER activity. Fe_3C-NCNFs afford a low OER onset potential (1.46 V vs. RHE). Remarkably, Fe_3C@NCNTs-NCNFs show an improved performance with the onset potential shifting to as low as 1.39 V vs. RHE shown in Fig. 6.7c [12].

Very interestingly, Peng et al. prepared sulfur modified $CaMnO_3$ nanotubes via electrospinning and subsequent calcination and sulfurization. Importantly, Fig. 6.8a depicts the SEM images of CMO/S. with uniform diameters of 250 nm. OER polarization curves of CMO/S-300 before and after 2000 cycles disclose their current decay about 3% presented in Fig. 6.8b. While Zn-air battery examination the CMO/S-300

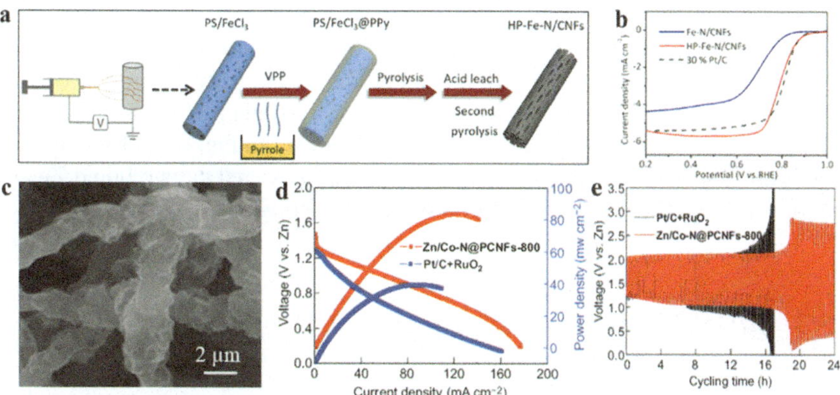

Fig. 6.6 **a** Schematic illustration of the preparation process of HP-Fe–N/CNFs. **b** RDE voltammograms of Fe-N-CNFs, HP-Fe-N/CNFs, and 30 wt% Pt/C. Reproduced from Ref. [24]. Copyright 2017, American Chemical Society. **c** SEM images of the a Zn/Co-ZIFsPCNFs-800. **d** Discharge polarization and power density curves of the Zn–air batteries using Zn/Co–N@PCNFs-800 and 20 wt% Pt/C+RuO_2 as ORR catalysts (mass loading of 1.2 mg cm^{-2}). **e** Galvanostatic discharge and charge cycling curves at 10 mA cm^{-2} with each cycle for 10 min (5 min charge and 5 min discharge) of rechargeable Zn-air batteries with the Zn/Co–N@PCNFs-800 and 20 wt% Pt/C+RuO_2 as the cathode catalyst. Reproduced from Ref. [27]. Copyright 2019. Open access

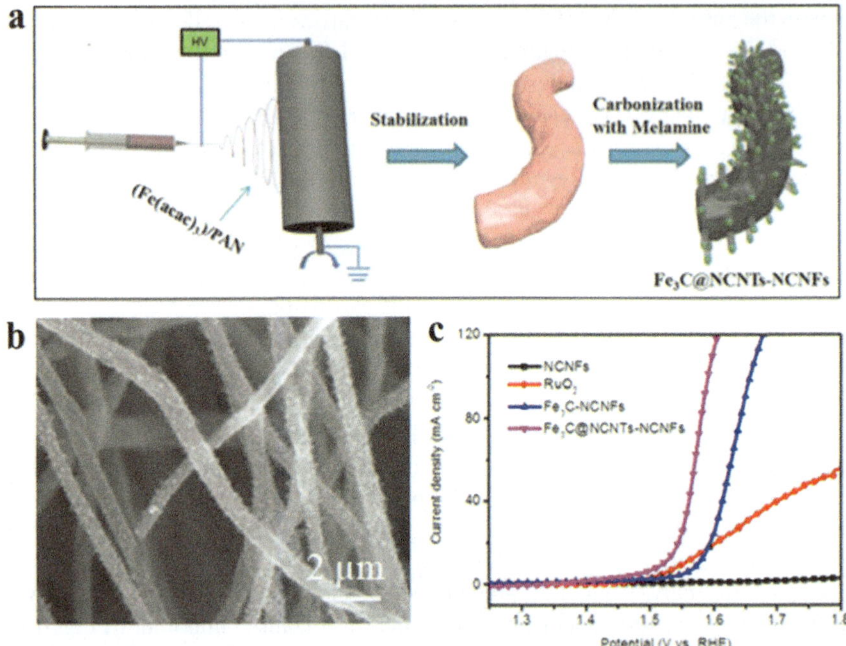

Fig. 6.7 **a** The fabrication process. **b** SEM images of Fe$_3$C@NCNTs-NCNFs. **c** OER catalytic properties of the as-prepared catalysts in 1 M KOH solution. Reproduced from Ref. [12]. Copyright 2017, The Royal Society of Chemistry

air–cathode generated an initial charge potential of 1.93 V and discharge potential of 1.25 V, with a small voltage gap of 0.67 V and a high round-trip efficiency of 64.8%. After 120 cycles, the CMO/S-300 air–cathode shows a slight performance loss with a small increase in the voltage gap by 0.13 V illustrated in Fig. 6.8c. The photo of an indicator light LED screen showing the "Zn-Air," powered by two liquid Zn–air batteries with the CMO/S air–cathode connected in series displayed in Fig. 6.8d.

6.5.3 *Bi/Multifunctional Electrocatalysts*

Oxygen reduction reaction (ORR) and oxygen evolution reaction (OER) are two key electrochemical processes in the field of metal-air batteries [34]. The ORR and OER participate multi-step proton-coupled electron transfer operation and the reversible electrochemical reaction of oxygen which takes to poor kinetics. Therefore, it is a vital necessity to develop bifunctional electrocatalysts [35, 36]. Noble metals Pt, IrO$_2$ and RuO$_2$ have been used as ORR and OER catalyst due to their excellent catalytic activity. However, the scarcity in the earth and high price hinders the mass-scale production [17, 21]. Hence, it is vital to develop efficient, durable and inexpensive electrocatalysts. As of now, a broad range of catalysts such as Co, Ni and Fe have been proposed as substitute electrocatalysts. Over the past year's development of highly active and stable electrocatalysts through simple electrospinning for both ORR and OER have been addressed. For example, Wang et al. reported Co@NS/CNTs-MCFs in which Co nanoparticles attached on carbon nanotubes-grafted multichannel carbon fibers, co-doped with nitrogen and sulfur. A prepared bifunctional catalyst provides exceptional stability through achieving 0.837 V for ORR, and about 362 mV overpotential at a current density of 10 mA cm^{-2} for OER. This remarkable performance ascribed to the hierarchical porous structure and multiple hetero atoms doping [37]. Likewise, Co nanoparticles [38], Co_3O_4 [39], $RuCoO_x$@Co/N-CNT [40] and $CuCo_2O_4$@C [41] have been used as bifunctional electrocatalysts which possess excellent Zn-air batteries in terms of displaying high charge–discharge profiles and cycle life. Among them, Wang et al. proposed rational design of $CuCo_2O_4$@C bifunctional depicts excellent ORR activity with an onset potential of 0.951 V and a low overpotential of 327 mV at 10 mA cm^{-2} for OER activity. Figure 6.9a represents SEM images of CCO@C which has tubular structure with abundant pores while charge discharge curves of CCO@C sample shows lower voltage gap of 0.79 V at 10 mAcm^{-2} see in Fig. 6.9b. It is stated that the rational design of surface structure through multiporous auxiliary and strong coupling with N-doped carbon [41]. Besides, Fe base catalysts like FeNi alloy [42], FeCo alloy [43], Fe_3C [44], Fe_3C/Fe [45], Fe/Co-N-C [46] have been accepted alternative bifunctional catalysts due to the low cost and efficient catalytic behaviour. Mainly, the FeCo alloy nanoparticle embedded with CNF exhibits web-like nanofibers see in Fig. 6.9c. The LSV curves demonstrate that FeCo-NCNFs-800 showed a smaller ΔE value (0.869 V) see in Fig. 6.9d, than other candidates due to the more active sites and mesoporous structure [43]. Although many catalysts available and reported some remarkable performance have been achieved by hybrid structure on the account of chemical and electrical properties when compared with the individual elements. Very recently, Ji et al. approached facile anchoring strategy to prepare Ni|MnO/CNF via electrospinning and subsequent calcination process. As prepared bifunctional catalysts SEM images shown in Fig. 6.9e. It can be observed that it has open mesoporous structure with 3D network. The Fig. 6.9f specify that the discharging voltage of Ni|MnO/CNF was higher than that of the Pt/C+RuO$_2$ cathode. In addition, Ni|MnO/CNF-ZAB also attained a high power density of 138.6 mW cm^{-2}, which was 124.6% of the value conveyed by the Pt/C+RuO$_2$ cathode, validating the great potential of Ni|MnO/CNF as a highly efficient air electrode that could replace expensive noble metal based materials in energy conversion and storage applications [47].

Another interesting approach is the preparation of CoNCNTF/CNFs carbon nanofibers supports nitrogen-doped, cobalt embedded carbon nanotubes/porous carbon flakes arrays hierarchical structures via free surface electrospinning method. The preparation strategy for CoNCNTF/CNF film is exemplified in Fig. 6.10a. At first, the free-surface electrospinning technique was established to construct polyacrylonitrile (PAN) nanofiber films in high output (Fig. 6.10a), signifying a hypothetically up-scalable industrial process. Secondly, a facile solution impregnation method was used to grow Co-based ZIFs on PAN nanofibers at room temperature.

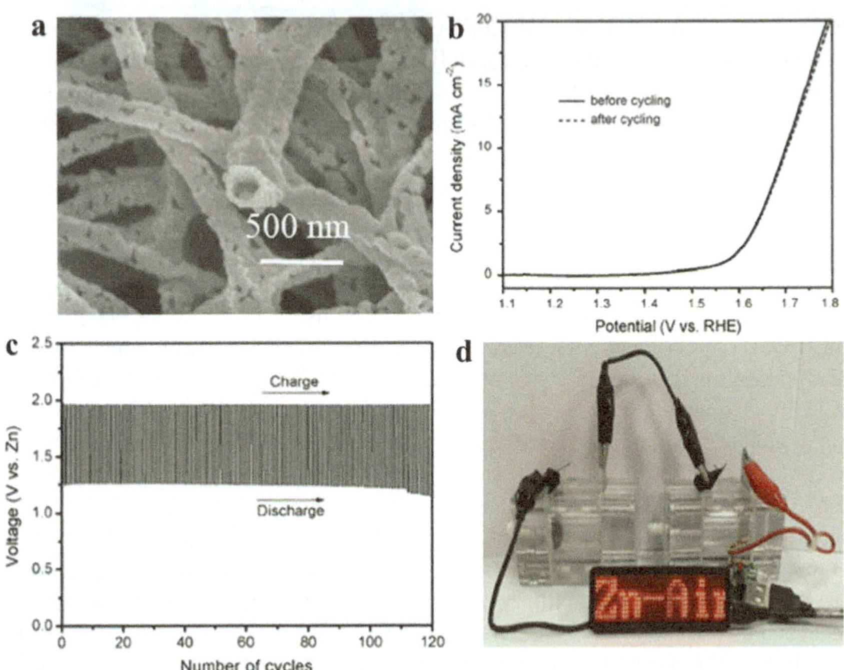

Fig. 6.8 **a** SEM images of CMO/S-300 OER polarization curves of CMO/S-300 before and after 2000 cycles. **b** OER polarization curves of CMO/S-300 before and after 2000 cycles. **c** Galvanostatic pulse cycling at 5 mA cm^{-2} with a duration of 400 s per cycle of CMO/S-300. **c, d** Photograph of an indicator light LED screen showing the "Zn-Air". Reproduced from Ref. [33]. Copyright 2018, Wiley–VCH

The typical FESEM image shown in Fig. 6.10b which has abundant tiny CNTs on the rough surface as a secondary nanostructure. Further, FETEM images represent that CNTs are composed of carbon shells with a thickness of ~2.4 nm (6–8 layers). The home mode solid-state ZABs display stable charge (~1.50 V) and discharge (~1.21 V) potentials at the current density of 0.5 mA cm^{-2} for 68 cycles (more than 11 h), which plays better than those of Pt/C-based air electrode illuminated in Fig. 6.10c. The remarkable battery performance can be accredited to the following reason of the fast mass-transport, flexibility. Good conductivity and promising bifunctional catalytic actions boost the ORR/OER [48].

Besides, the bifunctional electrocatalysts (Co$_3$O$_4$/Mn$_3$O$_4$/CN$_x$@CNFs) consequent from Co, Mn-ZIFs were first prepared by electrospinning combined with pyrolysis and oxidation process which is demonstrated in schematic illustration see in Fig. 6.11a. The surface morphology of each step provides clear idea of growth mechanism. The discharge and charge polarization curves of a zinc battery using Co$_3$O$_4$/Mn$_3$O$_4$/CN$_x$@CNFs as bifunctional catalyst compared with Pt/C+IrO$_2$ shown in see in Fig. 6.11b. The initial charge potential, discharge potential and voltage gap of the zinc-air battery made of Co$_3$O$_4$/Mn$_3$O$_4$/CN$_x$@CNFs as air

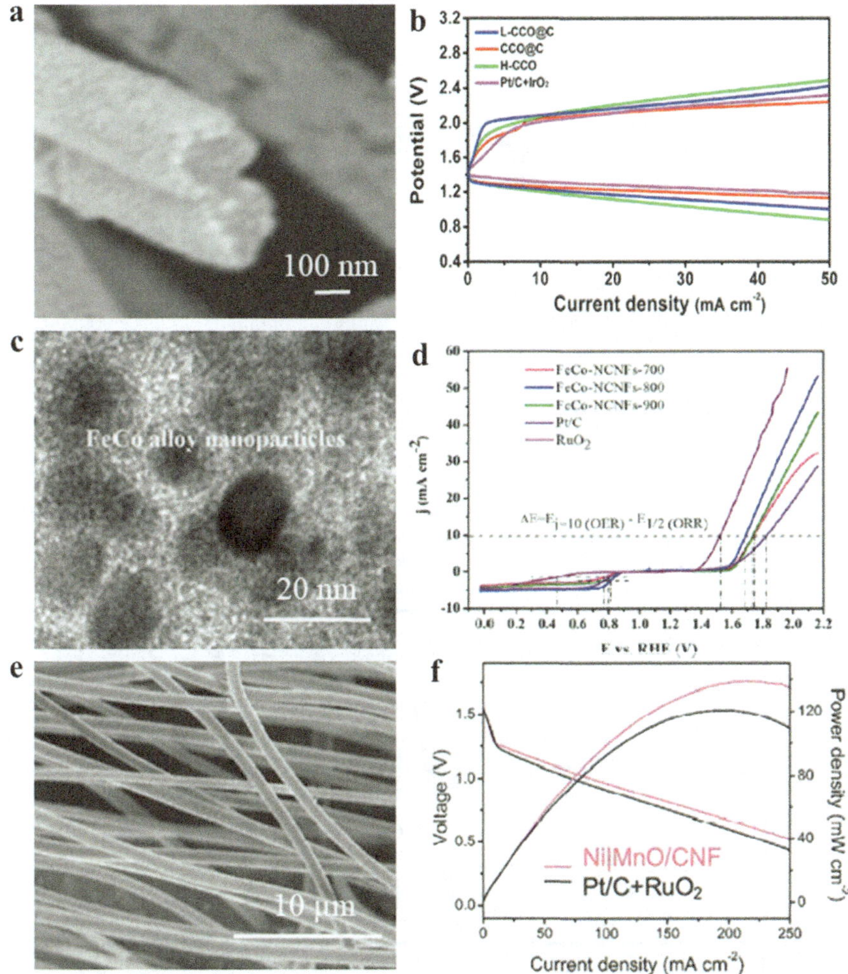

Fig. 6.9 **a** SEM images of CCO@C. **b** Charge and discharge polarization curves of LCCO@C, CCO@C, H-CCO, and Pt/C+IrO$_2$ samples. Reproduced from Ref. [41]. Copyright 2017, American Chemical Society. **c** TEM images of FeCo-NCNFs-800. **d** LSV curves of different catalysts showing the bifunctional ORR/OER activities in 0.1 M KOH at 1600 rpm. Reproduced from Ref. [43]. Copyright 2019, American Chemical Society. **e** SEM image of Ni/MnO/CNF catalyst. **f** Discharge polarization curves and corresponding power density plots for Ni|MnO/CNF-based and Pt/C+RuO$_2$-based ZABs. Reproduced from Ref. [47]. Copyright 2020 Wiley–VCH

electrodes were 2.19 V, 1.03 V and 1.16 V, respectively whereas over 50 h, the Co$_3$O$_4$/Mn$_3$O$_4$/CN$_x$@CNFs displays an insignificant property decay [49], at the same time, the Pt/C voltage gap increases expressively(Fig. 6.11b). Besides, Table 6.1 reveals some notable examples of electrospun based catalyst and their significant performance for zinc-air batteries.

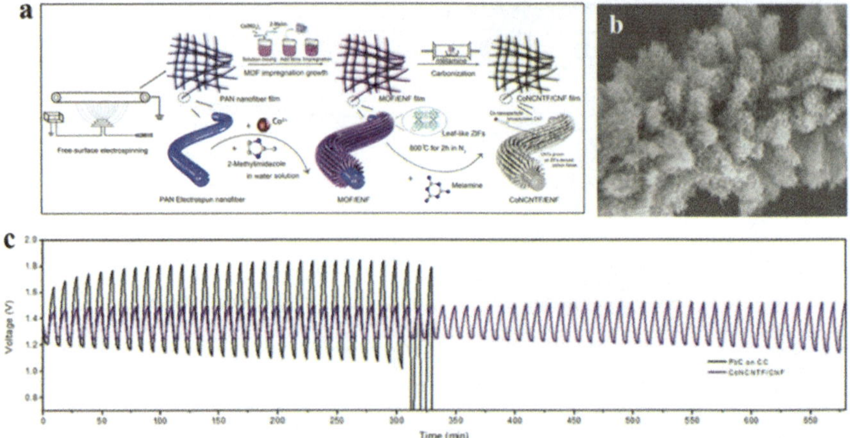

Fig. 6.10 **a** Schematic illustration of the fabrication process of CoNCNTF/CNF film typical. **b** FESEM images of CoNCNTF/CNF. **c** Galvanostatic discharge–charge cycling curves of the flexible all-solid-state ZABs with CoNCNTF/CNF film and Pt/C coated carbon cloth as flexible air electrode. Reproduced from Ref. [48]. Copyright 2018, Elsevier

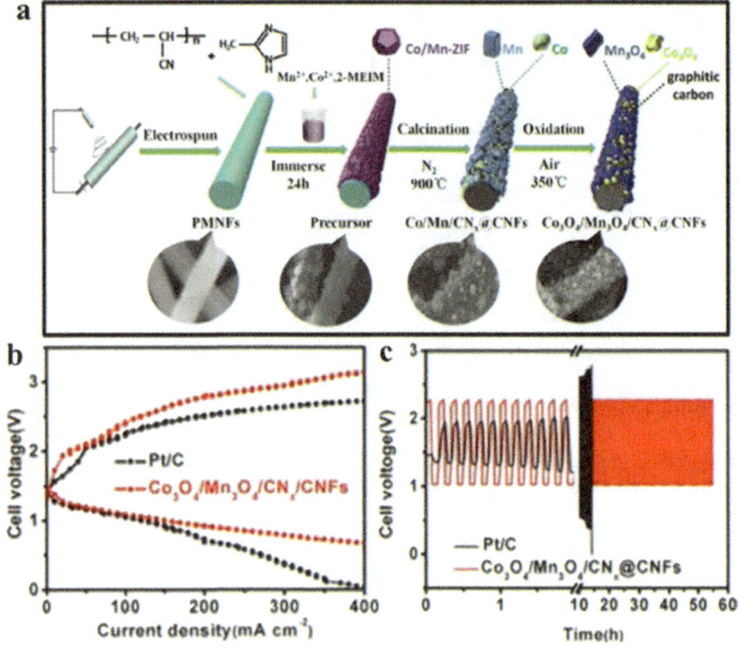

Fig. 6.11 **a** Schematic illustration of the preparation process of Co_3O_4, Mn_3O_4 and CN_x supported on 1D carbon nanofibers. **b** Charge and discharge polarization curves of Zn-air batteries using $Co_3O_4/Mn_3O_4/CN_x$@CNFs or Pt/C-IrO_2 as both air electrodes. **c** Galvanostatic charge/discharge test of the home-made ZnABs built with $Co_3O_4/Mn_3O_4/CNx$@CNFs catalyst at 5 mA cm^2. Reproduced from Ref. [49]. Copyright 2018 Elsevier

Table 6.1 Different kind of electrospun electrocatalysts materials for Zn-Air batteries

Materials	Precursor	Process	Zn-Air performance	References
CNCFs	PAN/Co(Ac)$_2$	700, 800 and 900 °C 3 h Ar:/H2 (95%: (5%)	Power density 51.7 mW cm^{-2}	[38]
FeCo-NCNFs-800	PAN/DMF Co$_3$[Fe(CN)$_6$]$_2$ · nH$_2$O	700, 800, and 900 °C for 2 h in Ar	Power density (74 mW cm^{-2}	[43]
Fe/N-HCNFs	PAN/ZIF-8/FeCl3	1st 260 °C for 2 h in N$_2$ 2nd 900 °C for 2 h in N$_2$	Power density of 50 mW cm^{-2}	[45]
NCNF film	ODA/BPDA/DMAc	1st 350 °C for 1 h in Ar 2nd (900, 1000 and 1100 °C) for 1 h in Ar	Power density of 185 mW cm^{-2}	[34]
Co SA@NCF/CNF	PAN/ Co-ZIF-Ls	850 °C for 2.5 h	Battery capacity of 530.17 mAh g$_{zn}^{-1}$	[9]
CoNCNTF/CNFs	Co(NO$_3$)$_2$.6H2O MeIM/PAN/DMF	800 °C for 2 h	Power density of 63 mWh cm^{-3}	[50]
Fe-NHCFs	Zn, Fe-ZIF/PAN/DMF	1st 250 °C for 2 h in N$_2$ 2nd 900 °C for 2 h in N$_2$	Power density of 61 mW cm^{-2}	[15]
FeCo@MNC	FeCl$_3$ · 6H$_2$O, 2,2-bipyridine/Co(NO$_3$)$_2$ · 6H$_2$O/DMF/PVP	1st 280 °C for 1 h in air 2nd 900 °C for 2 h in N$_2$	Power density 115 mW cm^{-2}	[46]
Co$_3$O$_4$/Mn$_3$O$_4$/CNx@CNFs	Co(NO$_3$)$_2$ · 6H$_2$O Mn(Ac)$_2$ · 4H$_2$O PAN/MeIM/DMF	900 °C for 1 h in N$_2$	Power density 265 mW cm^{-2}	[49]

References

1. Y. Li, H. Dai, Recent advances in zinc–air batteries. Chem. Soc. Rev. **43**(15), 5257–5275 (2014)
2. J. Fu, Z.P. Cano, M.G. Park, A. Yu, M. Fowler, Z. Chen, Electrically rechargeable zinc–air batteries: progress, challenges, and perspectives. Adv. Mater. **29**(7), 1604685 (2017)
3. J.S. Lee, S. Tai Kim, R. Cao, N.S. Choi, M. Liu, K.T. Lee, J. Cho, Metal–air batteries with high energy density: Li–air versus Zn–air. Adv. Energy Mater. **1**(1), 34–50 (2011)
4. X.G. Zhang, Fibrous zinc anodes for high power batteries. J. Power Sources **163**(1), 591–597 (2006)
5. T.Y. Ma, S. Dai, S.Z. Qiao, Self-supported electrocatalysts for advanced energy conversion processes. Mater. Today **19**(5), 265–273 (2016)
6. D.U. Lee, J.Y. Choi, K. Feng, H.W. Park, Z. Chen, Advanced extremely durable 3D bifunctional air electrodes for rechargeable zinc-air batteries. Adv. Energy Mater. **4**(6), 1301389 (2014)
7. X. Chen, B. Liu, C. Zhong, Z. Liu, J. Liu, L. Ma, Y. Deng, X. Han, T. Wu, W. Hu, Ultrathin Co3O4 layers with large contact area on carbon fibers as high-performance electrode for flexible zinc–air battery integrated with flexible display. Adv. Energy Mater. **7**(18), 1700779 (2017)
8. B. Li, X. Ge, F.T. Goh, T.A. Hor, D. Geng, G. Du, Z. Liu, J. Zhang, X. Liu, Y. Zong, Co3O4 nanoparticles decorated carbon nanofiber mat as binder-free air-cathode for high performance rechargeable zinc-air batteries. Nanoscale **7**(5), 1830–1838 (2015)
9. D. Ji, L. Fan, L. Li, S. Peng, D. Yu, J. Song, S. Ramakrishna, S. Guo, Atomically transition metals on self-supported porous carbon flake arrays as binder-free air cathode for wearable Zinc−Air batteries. Adv. Mater. **31**(16), 1808267 (2019)
10. S. Jiang, J. Li, J. Fang, X. Wang, Fibrous-structured freestanding electrodes for oxygen electrocatalysis. Small 1903760 (2019)
11. F. Meng, H. Zhong, D. Bao, J. Yan, X. Zhang, In situ coupling of strung Co4N and intertwined N-C fibers toward free-standing bifunctional cathode for robust, efficient, and flexible Zn–air batteries. J. Am. Chem. Soc. **138**(32), 10226–10231 (2016)
12. Y. Zhao, J. Zhang, X. Guo, H. Fan, W. Wu, H. Liu, G. Wang, Fe3C@ nitrogen doped cnt arrays aligned on nitrogen functionalized carbon nanofibers as highly efficient catalysts for the oxygen evolution reaction. J. Mater. Chem. A **5**(37), 19672–19679 (2017)
13. Q. Guo, D. Zhao, S. Liu, S. Chen, M. Hanif, H. Hou, Free-standing nitrogen-doped carbon nanotubes at electrospun carbon nanofibers composite as an efficient electrocatalyst for oxygen reduction. Electrochim. Acta **138**, 318–324 (2014)
14. Q. Bai, F.C. Shen, S.L. Li, J. Liu, L.Z. Dong, Z.M. Wang, Y.Q. Lan, Cobalt@nitrogen-doped porous carbon fiber derived from the electrospun fiber of bimetal-organic framework for highly active oxygen reduction. Small Methods **2**(12), 1800049 (2018)
15. M. Wu, C. Li, R. Liu, Freestanding 1D hierarchical porous Fe-N-doped carbon nanofibers as efficient oxygen reduction catalysts for Zn–Air batteries. Energy Technol. **7**(3), 1800790 (2019)
16. B. Li, K. Igawa, J. Chai, Y. Chen, Y. Wang, D.W. Fam, N.N. Tham, T. An, T. Konno, A. Sng, String of pyrolyzed ZIF-67 particles on carbon fibers for high-performance electrocatalysis. Energy Storage Mater. **25**, 137–144 (2020)
17. S. Zhao, H. Yin, L. Du, L. He, K. Zhao, L. Chang, G. Yin, H. Zhao, S. Liu, Z. Tang, Carbonized nanoscale metal–organic frameworks as high performance electrocatalyst for oxygen reduction reaction. ACS Nano **8**(12), 12660–12668 (2014)
18. Y. Qiu, J. Yu, T. Shi, X. Zhou, X. Bai, J.Y. Huang, Nitrogen-doped ultrathin carbon nanofibers derived from electrospinning: large-scale production, unique structure, and application as electrocatalysts for oxygen reduction. J. Power Sources **196**(23), 9862–9867 (2011)

19. K. Liang, L. Wang, Y. Xu, Y. Fang, Y. Fang, W. Xia, Y.-N. Liu, Carbon dots self-decorated heteroatom-doped porous carbon with superior electrocatalytic activity for oxygen reduction. Electrochim. Acta 135666 (2020)
20. G.S. Park, J.-S. Lee, S.T. Kim, S. Park, J. Cho, Porous nitrogen doped carbon fiber with churros morphology derived from electrospun bicomponent polymer as highly efficient electrocatalyst for Zn–air batteries. J. Power Sources **243**, 267–273 (2013)
21. X. Yan, Y. Jia, J. Chen, Z. Zhu, X. Yao, Defective-activated-carbon-supported Mn–Co nanoparticles as a highly efficient electrocatalyst for oxygen reduction. Adv. Mater. **28**(39), 8771–8778 (2016)
22. X. Lu, C. Wang, Y. Wei, One-dimensional composite nanomaterials: synthesis by electrospinning and their applications. Small **5**(21), 2349–2370 (2009)
23. X. Zhou, Y. Qiu, J. Yu, J. Yin, S. Gao, Tungsten carbide nanofibers prepared by electrospinning with high electrocatalytic activity for oxygen reduction. Int. J. Hydrogen Energy **36**(13), 7398–7404 (2011)
24. Y. Zhao, Q. Lai, Y. Wang, J. Zhu, Y. Liang, Interconnected hierarchically porous Fe, N-codoped carbon nanofibers as efficient oxygen reduction catalysts for Zn–Air batteries. ACS Appl. Mater. Interfaces **9**(19), 16178–16186 (2017)
25. J. Guo, B. Chen, Q. Hao, J. Nie, G. Ma, Pod-like structured Co/CoOx nitrogen-doped carbon fibers as efficient oxygen reduction reaction electrocatalysts for Zn-air battery. Appl. Surf. Sci. **456**, 959–966 (2018)
26. Z. Li, B. Chen, X. Wang, J. Nie, G. Ma, Electrospun bamboo-like Fe3C encapsulated Fe-Si-N co-doped nanofibers for efficient oxygen reduction. J. Colloid Interface Sci. **546**, 231–239 (2019)
27. Q. Niu, B. Chen, J. Guo, J. Nie, X. Guo, G. Ma, Flexible, porous, and metal–heteroatom-doped carbon nanofibers as efficient ORR electrocatalysts for Zn–air battery. Nano-Micro Lett. **11**(1), 8 (2019)
28. C. Wang, Q. Li, J. Guo, Y. Ren, J. Zhang, F. Yan, Metal-containing Ionic Liquid/Polyacrylonitrile-derived carbon nanofibers for oxygen reduction reaction and flexible Zn–Air battery. Chem. Asian J. **14**(11), 2008–2017 (2019)
29. C. Busacca, S. Zignani, A. Di Blasi, O. Di Blasi, M.L. Faro, V. Antonucci, A. Aricò, Electrospun NiMn2O4 and NiCo2O4 spinel oxides supported on carbon nanofibers as electrocatalysts for the oxygen evolution reaction in an anion exchange membrane-based electrolysis cell. Int. J. Hydrogen Energy **44**(38), 20987–20996 (2019)
30. S. Moon, Y.-B. Cho, A. Yu, M.H. Kim, C. Lee, Y. Lee, Single-step electrospun Ir/IrO2 nanofibrous structures decorated with au nanoparticles for highly catalytic oxygen evolution reaction. ACS Appl. Mater. Interfaces **11**(2), 1979–1987 (2018)
31. V.D. Silva, T.A. Simões, F.J. Loureiro, D.P. Fagg, F.M. Figueiredo, E.S. Medeiros, D.A. Macedo, Solution blow spun nickel oxide/carbon nanocomposite hollow fibres as an efficient oxygen evolution reaction electrocatalyst. Int. J. Hydrogen Energy **44**(29), 14877–14888 (2019)
32. A. Morinaga, H. Tsutsumi, Y. Katayama, Electrospun Cu-deposited flexible fibers as an efficient oxygen evolution reaction electrocatalyst. ChemPhysChem **20**(22), 2973–2980 (2019)
33. S. Peng, X. Han, L. Li, S. Chou, D. Ji, H. Huang, Y. Du, J. Liu, S. Ramakrishna, Electronic and defective engineering of electrospun CaMnO3 nanotubes for enhanced oxygen electrocatalysis in rechargeable zinc–air batteries. Adv. Energy Mater. **8**(22), 1800612 (2018)
34. Q. Liu, Y. Wang, L. Dai, J. Yao, Scalable fabrication of nanoporous carbon fiber films as bifunctional catalytic electrodes for flexible Zn-Air batteries. Adv. Mater. **28**(15), 3000–3006 (2016)

35. D. Zhu, Q. Zhao, G. Fan, S. Zhao, L. Wang, F. Li, J. Chen, Photoinduced oxygen reduction reaction boosts the output voltage of a zinc-air battery. Angew. Chem. Int. Ed. **58**(36), 12460–12464 (2019)
36. I.S. Amiinu, X. Liu, Z. Pu, W. Li, Q. Li, J. Zhang, H. Tang, H. Zhang, S. Mu, From 3D ZIF nanocrystals to Co–Nx/C nanorod array electrocatalysts for ORR, OER, and Zn–air batteries. Adv. Func. Mater. **28**(5), 1704638 (2018)
37. Z. Wang, S. Peng, Y. Hu, L. Li, T. Yan, G. Yang, D. Ji, M. Srinivasan, Z. Pan, S. Ramakrishna, Cobalt nanoparticles encapsulated in carbon nanotube-grafted nitrogen and sulfur co-doped multichannel carbon fibers as efficient bifunctional oxygen electrocatalysts. J. Mater. Chem. A **5**(10), 4949–4961 (2017)
38. H. Li, M. An, Y. Zhao, S. Pi, C. Li, W. Sun, H.-G. Wang, Co nanoparticles encapsulated in N-doped carbon nanofibers as bifunctional catalysts for rechargeable Zn-air battery. Appl. Surf. Sci. **478**, 560–566 (2019)
39. L. Qiu, X. Han, Q. Lu, J. Zhao, Y. Wang, Z. Chen, C. Zhong, W. Hu, Y. Deng, Co3O4 nanoparticles supported on N-doped electrospinning carbon nanofibers as an efficient and bifunctional oxygen electrocatalyst for rechargeable Zn–air batteries. Inorganic Chem. Front. **6**(12), 3554–3561 (2019)
40. J. Yang, L. Chang, H. Guo, J. Sun, J. Xu, F. Xiang, Y. Zhang, Z. Wang, L. Wang, F. Hao, Electronic structure modulation of bifunctional oxygen catalysts for rechargeable Zn–air batteries. J. Mater. Chem. A (2020)
41. X. Wang, Y. Li, T. Jin, J. Meng, L. Jiao, M. Zhu, J. Chen, Electrospun thin-walled CuCo2O4@C nanotubes as bifunctional oxygen electrocatalysts for rechargeable Zn–air batteries. Nano Lett. **17**(12), 7989–7994 (2017)
42. Z. Wang, J. Ang, J. Liu, X.Y.D. Ma, J. Kong, Y. Zhang, T. Yan, X. Lu, FeNi alloys encapsulated in N-doped CNTs-tangled porous carbon fibers as highly efficient and durable bifunctional oxygen electrocatalyst for rechargeable zinc-air battery. Appl. Catal. B **263**, 118344 (2020)
43. L. Yang, S. Feng, G. Xu, B. Wei, L. Zhang, Electrospun MOF-based FeCo nanoparticles embedded in nitrogen-doped mesoporous carbon nanofibers as an efficient bifunctional catalyst for oxygen reduction and oxygen evolution reactions in zinc-air batteries. ACS Sustain. Chem. Eng. **7**(5), 5462–5475 (2019)
44. G. Ren, X. Lu, Y. Li, Y. Zhu, L. Dai, L. Jiang, Porous core–shell Fe3C embedded N-doped carbon nanofibers as an effective electrocatalysts for oxygen reduction reaction. ACS Appl. Mater. Interfaces **8**(6), 4118–4125 (2016)
45. Q. Li, J. Zhao, M. Wu, C. Li, L. Han, R. Liu, Hierarchical porous N-doped carbon nanofibers supported Fe3C/Fe nanoparticles as efficient oxygen electrocatalysts for Zn−Air batteries. Chem. Select **4**(2), 722–728 (2019)
46. C. Li, M. Wu, R. Liu, High-performance bifunctional oxygen electrocatalysts for zinc-air batteries over mesoporous Fe/Co-NC nanofibers with embedding FeCo alloy nanoparticles. Appl. Catal. B **244**, 150–158 (2019)
47. D. Ji, J. Sun, L. Tian, A. Chinnappan, T. Zhang, W.A.D.M. Jayathilaka, R. Gosh, C. Baskar, Q. Zhang, S. Ramakrishna, Engineering of the heterointerface of porous carbon nanofiber-supported nickel and manganese oxide nanoparticle for highly efficient bifunctional oxygen catalysis. Adv. Funct. Mater. 1910568 (2020)
48. D. Ji, L. Fan, L. Li, N. Mao, X. Qin, S. Peng, S. Ramakrishna, Hierarchical catalytic electrodes of cobalt-embedded carbon nanotube/carbon flakes arrays for flexible solid-state zinc-air batteries. Carbon **142**, 379–387 (2019)

49. L. Li, L. Fu, R. Wang, J. Sun, X. Li, C. Fu, L. Fang, W. Zhang, Cobalt, manganese zeolitic-imidazolate-framework-derived $Co_3O_4/Mn_3O_4/CN_x$ embedded in carbon nanofibers as an efficient bifunctional electrocatalyst for flexible Zn-air batteries. Electrochim. Acta 136145 (2020)
50. D. Ji, L. Fan, L. Li, N. Mao, X. Qin, S. Peng, S. Ramakrishna, Hierarchical catalytic electrodes of cobalt-embedded carbon nanotube/carbon flakes array for flexible solid-state zinc-air batteries. Carbon **142**, 379–387 (2019)

Chapter 7
Electrospinning of Nanofibers for Li-Air Battery

Abstract The mass scale electrical energy storage using rechargeable batteries sustains any future success in the global efforts to shift energy usage away from fossil fuels to renewable sources. Among the various rechargeable battery chemistries under development, the lithium-air battery is considered as the most promising future technology for long-range EVs (>500 km), because its theoretical energy density is about ten times higher. Over the last decade, extensive research on LAB technologies has been conducted. There are still many problems that need to be solved; however, including low capacity, poor rate capability, low round-trip efficiency. In this chapter, we first reviewed the basic working principle and cell structure. Afterwards, we demonstrated the summary of recent advancement of electrospun nanofibers, particularly binder-free and free-standing electrode microstructure and their electrocatalytic behavior when used as OER and ORR performance for high performance and flexible Li-air batteries.

Keywords Li-air battery · Nanostructured electrodes · Porous structure · Electrocatalysts

7.1 Li-Air Battery Working Principle and Cell Structure

Lithium-air battery (LAB) is regarded as the most promising future technology among the many rechargeable batteries in development, the for long-range electric vehicles, due to its theoretical energy density is about ten times higher than those of current LIBs. The LAB is usually arranged of a metallic Li anode (negative electrode) and a porous cathode (oxygen electrode positive electrode) with different types [1] of electrolytes provided in Fig. 7.1. There are four types of LAB, categorized according to the species of electrolyte: Li salts-dissolved in (i) non-aqueous (aprotic) solvents (ii) aqueous solvents (iii) hybrid (non-aqueous/aqueous) solvents, and (iv) all-solid-state electrolyte. Basically, LAB reaction mechanisms differ according to the electrolyte used. Li-air batteries use oxygen gas as a cathode material, and thus, porous carbon and catalyst composites must be added as the Li_2O_2 reservoir in the cathode [2].

S. Peng and P. R. Ilango, *Electrospinning of Nanofibers for Battery Applications*, https://doi.org/10.1007/978-981-15-1428-9_7

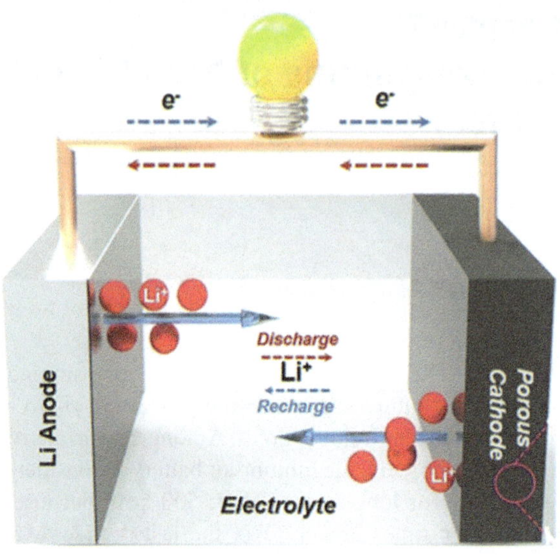

Fig. 7.1 Schematic representation of Li-Air battery. Reproduced from Ref. [1]. Copyright 2016, The Royal Society of Chemistry

The non-aqueous electrolyte system was first introduced by Abraham et al. and they recommended the subsequent stepwise reaction mechanism

$$2Li + O_2 \rightarrow Li_2O_2 \, E^0 = 3.10V \, vs \, Li/Li+ \tag{7.1}$$

$$4Li + O_2 \rightarrow 2Li_2O \, E^0 = 2.91V \, vs \, Li/Li+ \tag{7.2}$$

Standard cell potentials, E°, were calculated using the standard Gibbs free energy of formation. Then, the Bruce et al. proposed that Li_2O_2 is founded on charging and decomposes corresponding to the reaction, $Li_2O_2 \rightarrow O_2 + 2Li^+ + 2e^-$, based on ex-situ powder X-ray diffraction and in-situ mass spectroscopy [3]. After that, so many studies have been recommended to revise the theoretically reversible potentials of the above equations. Additionally, the basic effects of solvents on the ORR reveal the reaction mechanism of the oxygen electrode [4, 5].

7.2 Perspectives on Material Development

7.2.1 Anode

A lithium metal will be used as the anode which is supposed to accomplish a high energy density owing to its high electropositivity (-3.04 V vs standard hydrogen electrode) and the lightest metal (equivalent weight $= 6.94$ gmol^{-1}, specific gravity $= 0.53$ g cm^{-3}); it also has a remarkable specific capacity (3860 Ahkg^{-1}) [6]. Though, the grouping of Li metal and a liquid electrolyte solution is very challenging for rechargeable batteries since of the high reactivity of the active Li metal with any suitable polar aprotic solvent. Also, the SEI formation occurred due to the surface reaction between lithium metal and electrolyte species [7]. The surface morphology for Li deposition is dendritic and very porous, and the dendritic Li can be cut and separated from the anode substrate during the discharge. This isolated Li triggers a failure of anode materials and starts to a smaller cycle life. Consequently, dendritic Li formation and the electrochemical instability of a Li metal electrode persist major disputes that must be overcome to permit the function of Li-air batteries. The SEI layer on the Li electrode cannot suitably adjust the spectacular morphological changes of the Li metal upon Li deposition and dissolution due to the non-uniformity.

7.2.2 Cathode

The air electrode is made up of carbon, a catalyst, and a polymer binder and their design affect the electrochemical performance since reaction mechanisms of Li-air batteries, and that of fuel cells are same. In the case of hydrogen fuel cells, H^+ protons step through the electrolytes to connect oxide ions at the catalyst to form water whereas Li^+ step through the electrolytes and connect oxide ions at the air cathode surface to form a Li_2O_2 precipitate in the case of Li-air batteries. The LAB differs from H_2 fuel cell by two factors. Firstly, the reduction reaction of oxygen can occur without catalysts in LABs, but such non-catalyzed reactions are not possible for fuel cells. Secondly, the reduced product of oxygen in LABs is gathered as a solid in the pores of the cathode electrode. However, the reduced outcome of oxygen in the fuel cells is not collected because it forms water. Besides, it has some challenges like the pores of air cathodes in LABs become increasingly blocked by Li_2O_2 precipitates as the discharge proceeds. Thus pore size, electrode thickness, surface area wettability of electrolytes, and electrical contact of active materials to be considered in the fabrication of air electrode for LABs [2, 8–11]. Regarding the development catalysts for LABs, numerous reports have been published and stated the mechanism of implemented catalytic performance towards efficient and durable for future practical applications. For example, Bruce et al. have investigated the effect of various catalysts like Pt, $La_{0.8}Sr_{0.2}MnO_3$, Fe_2O_3, NiO, Fe_3O_4, Co_3O_4, CuO, and $CoFe_2O_4$ with a size of 1–5 μm [12]. As ORR and OER can take place even in the non-appearance of a catalyst, as stated. Still, catalysts can enable both reactions, while the discharge voltage is slightly altered by catalysts.

7.3 Binder Free Electrodes

LAB batteries can deliver a significantly high energy density of 3505 W h kg^{-1}, which will bring a start in operating range of EVs. Significant disputes persist for the practical applications of LABs, although the electrochemical performance has progressed remarkably recently by boosting the battery designs [13]. Recent reports say that the construction binder-free electrode or growing directly on the substrate displays a unique architecture which superbly enhances the catalytic performance and reduce the secondary reaction [14, 15]. Xue et al. proposed 3D hierarchical porous NCO@NCF hybrid film in which $NiCo_2O_4$ nanoparticles-decorated mesoporous N-doped carbon nanofibers by electrospinning method and their fabrication schematic diagram illustrated in Fig. 7.2a. During this experiment, the multi-step heat treatment was involved and finally achieved the binder-free film electrode. Figure 7.2b illustrates the full discharge and charge curves in the first cycle of the NCO@NCF cathode examined at a current density of 200 mA g^{-1}. The discharge capacity of the binder-free NCO@NCF cathode (5304 mAh g^{-1}) is much higher than that of the binder rich mashed NCO@NCF cathode (2920 mAh g^{-1}). In supplement, the binder-free NCO@NCF electrode presents low discharge and charge overpotentials of 0.22

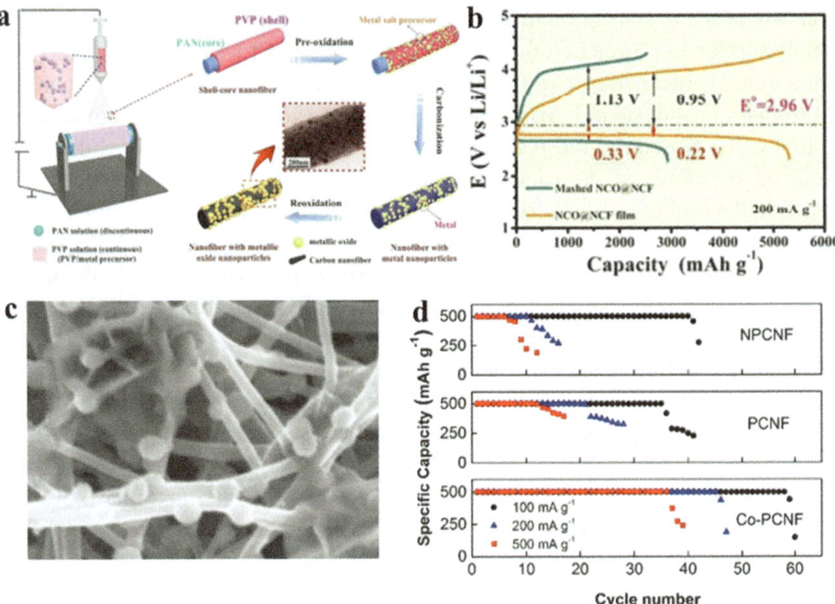

Fig. 7.2 **a** Illustrative diagram of the fabrication of self-supported NCO@NCF film. **b** The first full discharge/charge curves of film cathode. Reproduced from Ref. [16]. Copyright 2016, The Royal Society of Chemistry. **c** Co-PCNF cathodes at various after first discharge. **d** Rate performance of NPCNF, PCNF and Co-PCNF at various current densities. Reproduced from Ref. [17]. Copyright 2016, The Royal Society of Chemistry

and 0.95 V, respectively, which is advantageous for reaching a higher coulombic efficiency of ~100% in the charge process [16]. Later, Singhal et al. studied Co-PCNF in which porous carbon nanofibers decorated with cobalt nanoparticles with the high surface area. The SEM images of as-prepared Co-PCNF see in Fig. 7.2c revealed a toroid structure with 0.106 um diameter. When they explore their rate performance with coin type LABs, the electrode with Co-PCNF exhibits superior rate performance (Fig. 7.2d) when compare with other candidates like PCNF and NPCNF which clearly indicates that Co-PCNF has higher kinetics possible via pore formation [17].

7.4   Free-Standing Electrodes

With the intension to the concept of auspicious porous structures and evade the side reactions of carbon materials and adhesive. The free-standing cathode related catalysts become the actual and broadly used method. Mainly, cobalt-based free-standing catalysts have drawn wide notice as a electrode material for LABs [18, 19]. For example, carbon nanofibers based self-standing without binder a more graphitized mesoporous carbon layer with high electrical conductivity and surface area may be beneficial for Li-air battery application [20]. For instance, Cao et al. reported, free-standing Co/CNFs films through electrospinning method and described with the scheme as shown in Fig. 7.3a. The 3D hierarchical porous CNF with various amount of Co is synthesized by the electrospinning followed by thermal treatment. The average diameter of the as-spun fibers is less than 500 nm with microporous structure. Noted that micropores are dispersed both inside of and on the surface of the CNFs. Co-NPs are evenly distributed in the carbon fiber as suggested by the TEM image see in Fig. 7.3b. The charge–discharge profiles at different cycles of Co/CNF-7.4% cathode remained stable with the terminal charge voltage below 4.5 V as well as the average charge voltage of 4.1 V and discharge voltage of 2.58 V after 40 cycles Fig. 7.3c, subsequently, it can be detected that the discharge and charge curves are smooth and stable without any apparent oscillations [21].

Likewise, Co-doped and Co–Ni based free-standing electrode reported with remarkable LABs with the unique design of architecture [22, 23]. Diversely, activated carbon nanofibers based (ACNF) free-standing were formulated through electrospinning combining with CO_2 activation and then utilized for non-aqueous $Li–O_2$ battery cathodes which has porous structure with a rough surface. Obviously, the CNFs are homogeneously covered with Li-oxides established during the discharge shown in Fig. 7.4a. When exploring $Li–O_2$ cells with ACNF cathodes, it is noted that the charge terminal is only 4.4 V which is 200/300 mV lower than that of CNF/BP 2000 cathodes respectively depicted in Fig. 7.4b. The comparatively lower charge overpotential of ACNF cathode in this experiment may be associated to its 3D interconnected

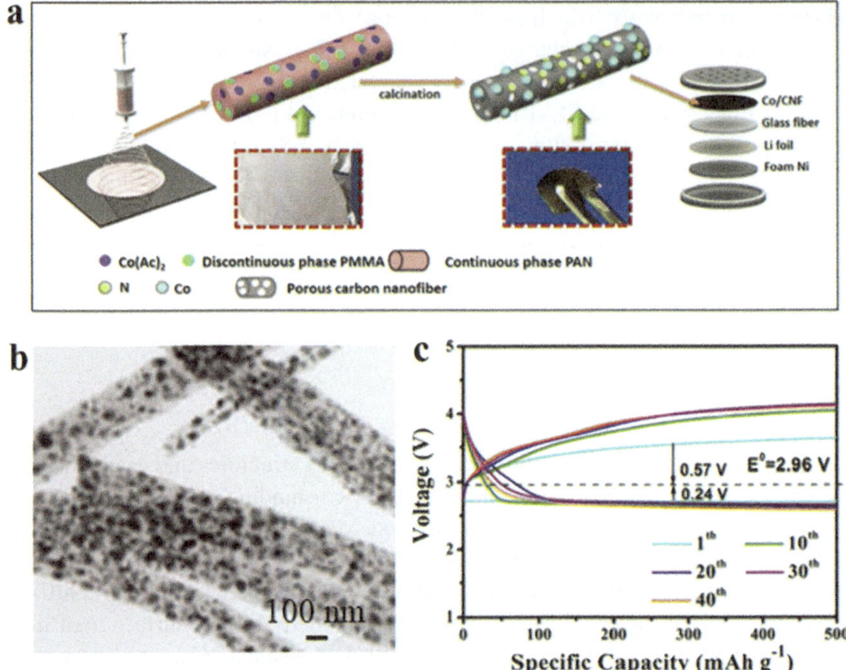

Fig. 7.3 **a** Schematic diagram of the preparation process of the free-standing Co/CNF hybrid catalyst. **b** TEM images of Co/CNF-7.4% cathode. **c** Discharge and charge profiles at different cycles of Co/CNF-7.4% cathode at 100 mA g^{-1}. Reproduced from Ref. [21]. Copyright 2017, Elsevier

microstructure, and the unique flake-like structure of Li_2O_2 and the following cycle performance reveals as prepared cathode remains higher than 2.0 V until 51 [24] cycles for ACNF while CNF reveal lower than that even after 19 cycles displayed in Fig. 7.4c, d.

7.5   Other Catalysts

7.5.1   Metal Oxides Electrocatalysts

In the viewpoint of nanostructured metal oxide-based catalysts with supporting of carbon, have been recently recommended as ORR and OER catalysts in Li–O$_2$ batteries due to many aspects including the availability of more active sites, a

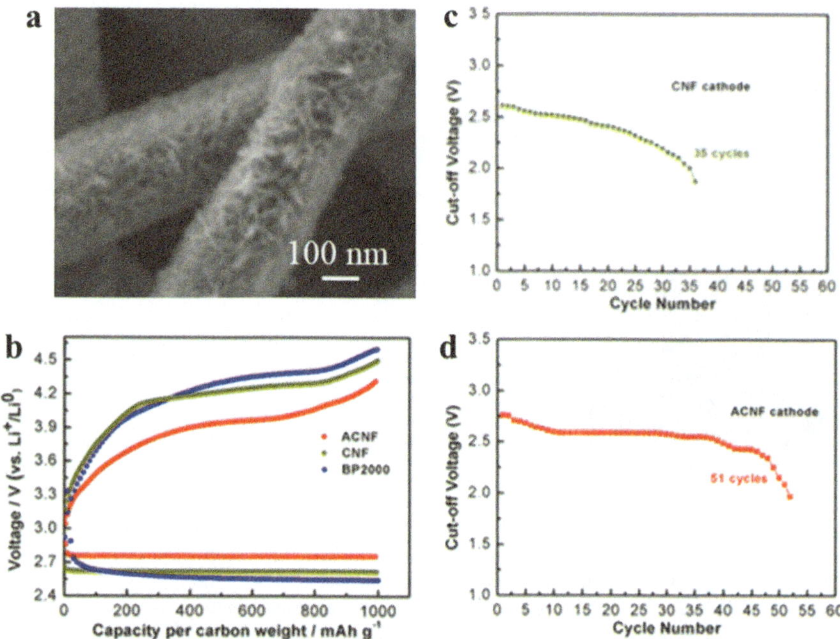

Fig. 7.4 **a** SEM images of ACNF cathodes. **b** First charge–discharge curves of Li–O_2 cells at a current density of 200 mA g^{-1}. **c, d** Cycle performance of CNF and ACNF cathode for Li–O_2 batteries. Reproduced from Ref. [24]. Copyright 2015, American Chemical Society

high surface area fast diffusion pathway Li_2O_2 phases form and decompose preferentially and excellent catalytic activities [25–27]. Ryu et al. investigated Co_3O_4 NF/GNF composite as a bifunctional catalyst and their preparation outline provided in Fig. 7.5a. It can be noted that a large amount of Co_3O_4 NFs, which randomly cross each other, were uniformly connected on the RGO nanoflakes Fig. 7.5b.

Intriguingly, the distinctive combination of the Co_3O_4 NF/GNF exhibited a high discharge capacity of 10,500 mAh/g and stable cyclability for 80 cycles under the specific capacity limit, compared to Co_3O_4 NP, Co_3O_4 NF, and Co_3O_4 NF/RGO which strongly attributed to improved bifunctional catalytic activity in both the OER and ORR [28]. Also, the similar idea proposed by another group with MnO_3O_4/C composite delivers superior ORR activity as proven by an onset potential as high as 0.9 V versus RHE and a higher half-wave potential (0.8 V vs. RHE) see Fig. 7.5c. Moreover, the charge–discharge profile indicates reduced discharge–charge overpotentials when using Mn_3O_4/C nanofiber catalyst [29] provided in Fig. 7.5d.

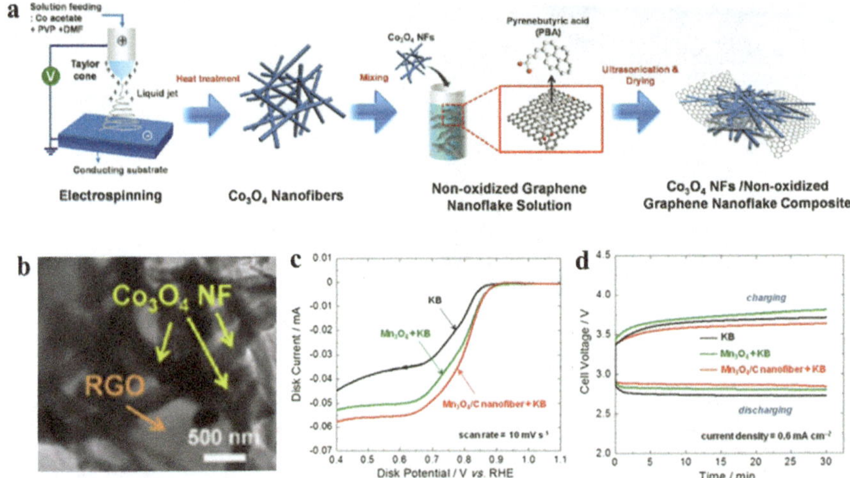

Fig. 7.5 **a** Schematic illustration of the synthetic strategy of the Co$_3$O$_4$ NF/GNF composite. **b** TEM images of Co$_3$O$_4$ NF/RGO composite. Reproduced from Ref. [28]. Copyright 2013, American Chemical Society. **c** ORR polarization curves on the RDE at 1200 rpm in 0.1 M KOH for [KB], [Mn$_3$O$_4$ + KB] and [Mn$_3$O$_4$/C nanofiber + KB]. **d** Discharge and charge profiles of the hybrid Li–O$_2$ batteries at a current density of 0.6 mA cm^{-2}. Reproduced from Ref. [29]. Copyright 2012, The Royal Society of Chemistry

7.5.2 *Metal and Other Catalysts*

Over the past few years, so many catalysts including carbon materials, metal nitrides metal composites etc. have been explored to implement as an effective catalyst future important scientific and modern applications, particularly lithium-air batteries [30, 31]. Firstly, metal-based catalysts like Au and Pt, which are exceedingly active for the ORR, with another metal (Pt) that is highly active for the OER [32]. Previously, mesoporous carbon nanofibers doped with Palladium nanoparticles (Pd CNFs) were produced by electrospinning with consequent thermal treatment processes and used as electrocatalysts at the oxygen cathode of the Li–O$_2$ battery. As prepared Pd CNFs reveals Pd decorated on the 300 to 500 nm with tens of micrometers in the length of CNFs shown in Fig. 7.6a. Comparative galvanostatic charge–discharge studies profile shown in Fig. 7.5b. Undoubtedly, at the first cycle, in the presence of Pd (Pd2.5, Pd2.5A and Pd5) the oxidation potential stay under 3.8 V, whereas it increased above 4.25 V in its absence. Remarkably low charge potential of 3.2 V was observed for Pd2.5A [33] illustrated in Fig. 7.6b.

Besides, carbon nanofibers (CNFs) were synthesized via simple electrospinning methods with admiration to electrospinning time with polyacrylonitrile (PAN). As

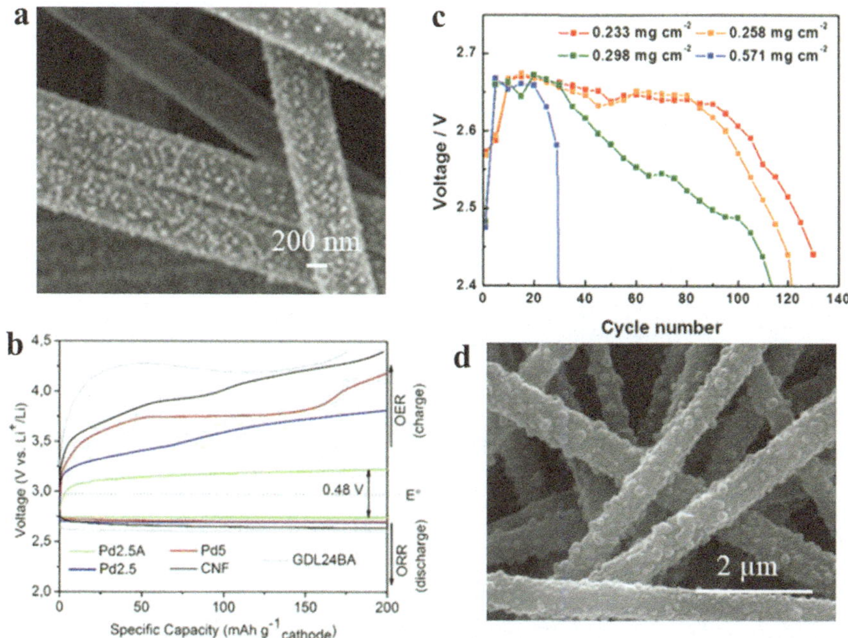

Fig. 7.6 a SEM images of Pd$_{2.5}$. **b** Galvanostatic discharge and charge of the Li–O$_2$ cells at the curtailed capacity of 200 mAh g^{-1} at the applied current of 20 mA g^{-1} with Pd based cathode. Reproduced from Ref. [33]. Copyright 2016, The Royal Society of Chemistry. **c** Voltage profiles for discharging at the loading amount and cycle number. **d** SEM images of CNFs with loading levels of 0.571 mg cm^{-2} after the charge–discharge measurement. Reproduced from Ref. [34]. Copyright 2017, American Scientific Publisher

well, carbonization was completed to convert the polymers to carbon materials. The willing CNF sheets were measured to be 0.233, 0.258, 0.295, and 0.571 mg cm^{-2}, respectively The terminal discharge voltage was monitored by cycle number as seen in Fig. 7.6d. Both the CNFs with loadings of 0.233 mg cm^{-2} and 0.258 mg cm^{-2} maintained a stable discharge voltage above 2.65 V for 100 cycles. On the other hand, the CNFs with high degrees of loading endlessly displayed a decreasing trend of discharge potential in behaviour with an increase in charge–discharge cycles and the SEM image of CNF after charge–discharge testing were examined and revealed discharge products but still maintains pore structure [34] shown in Fig. 7.6d.

Later, Ji et al. attempted novel strategy to mass integrate nonprecious transition-metal-based nitrogen/oxygen co-doped carbon nanotubes (CNTs) grown on carbon-nanofiber films (MNO-CNT-CNFFs, M = Fe, Co, Ni) via a facile free-surface electro-spinning technique followed by in situ growth carbonization. The complete synthesis procedure clearly demonstrated in Fig. 7.7a.

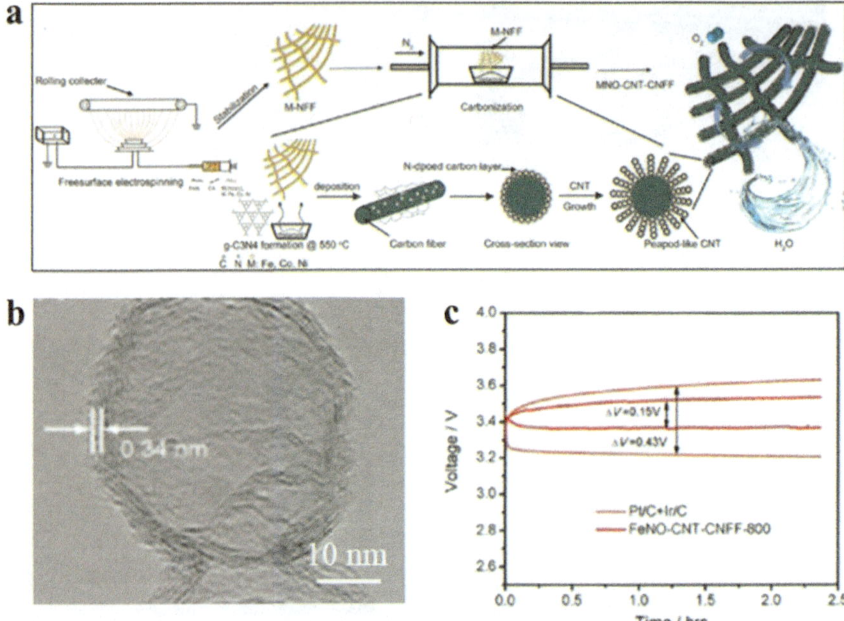

Fig. 7.7 **a** Schematic diagram of synthesis strategy. **b** TEM images of FeNO-CNT-CNFF.
c Comparison of voltage gap between charge–discharge voltage plateaus of hybrid Li-air batteries
with different catalysts. Reproduced from Ref. [35]. Copyright 2017, American Chemical Society

To get insights, as prepared electrospun carbon fibers with diameter of $\sim$600 nm
were evenly covered by the in situ grown Fe-based CNTs which had peapod-
like structures. These nanoscale peapods possess diameters of approximately 10–
30 nm and lengths in the range of several hundred nanometers. The "peas" were
well-crystallized, 5–20 nm diameter Fe-based particles having a lattice distance of
$\sim$0.21 nm (Fig. 7.7b), and were enclosed in "peapods" comprising 4–6 graphitic
layers and the first charge–discharge curves of the hybrid cell with different cathodes, where ΔV denotes the voltage difference between the charge and discharge
voltages. Effectively, the FeNO-CNT-CNFF-800-based battery exhibits the excellent performance in the open-air test with a ΔV of 0.15 V at a current density of
0.03 mA cm^{-2}, much lower than that of the Pt/C + Ir/C air electrode (ΔV = 0.43 V)
[35] (Fig. 7.7c) that ascribed to the hierarchically porous structure with hydrophilic
surface allowing fast mass-transport of oxygen and electrolyte, and the excellent
conductivity of CNTs and interconnected 1D carbon fiber without binder providing
a "highway" for electron transfer. Meanwhile, Li et al. produced Fe/Fe$_3$C carbon
nanofibers as a bifunctional catalyst for Li–O$_2$ batteries. As developed Fe/Fe$_3$C
CNF are involved in randomly oriented, overlapped, continuous and interconnected
nanofibers with the average diameter of about 350 nm. The corresponding TEM,
HR-TEM and SAED (Fig. 7.8a–c) pattern confirms the Fe$_3$C planes (111) with an
interlayer spacing of 0.30 nm.

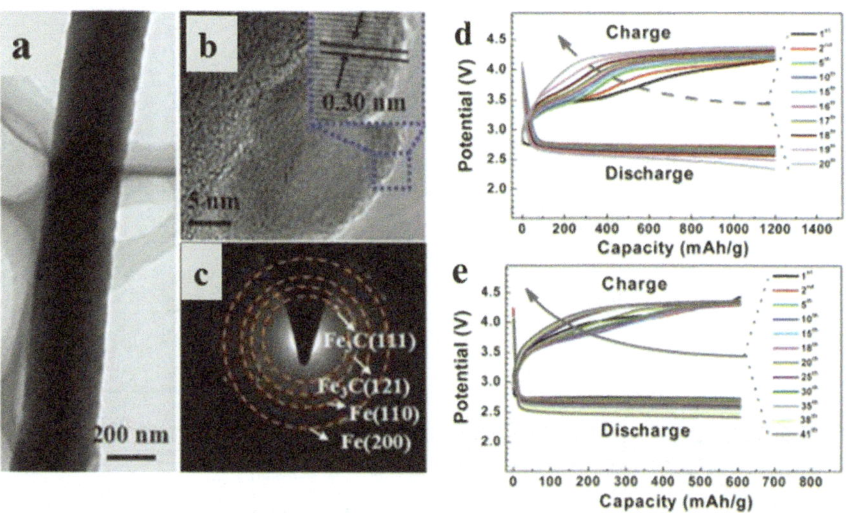

Fig. 7.8 a–c The TEM, HR-TEM and the corresponding SAED pattern of Fe/Fe$_3$C-CNFs. **d, e** Cycling performance of the Fe/Fe$_3$C–CNF electrodes at 300 mA g^{-1} with a restricting capacity of 1200 mA h g^{-1} and 600 mA h $^{-1}$. Reproduced from Ref. [36]. Copyright 2014, The Royal Society of Chemistry

Following the cycle stability of Fe/Fe$_3$C-CNF cathodes with different restricted capacities was examined. The cycling performance for the Fe/Fe$_3$C-CNF cathode at 300 mA g^{-1} with a restricting capacity of 1200 mAh g^{-1} see in Fig. 7.8d. It is found that their D-C plateaus for the first cycle were about 2.70 V and 3.75 V, respectively, exposing its reasonable round-trip efficiency and good redox reaction for the ORR and OER process. Though, the D-C potential difference increased in the following 19 cycles, suggesting a progressively improved polarization. To further prove the excellent cycle performance of the Fe/Fe$_3$C-CNF cathode, LOB was cycled with a lower restricting capacity of 600 mAh g^{-1} also at a large current density of 300 mA g^{-1}. As predicted, the Fe/Fe$_3$C–CNF cathode exhibited exceptional cycle stability up to 40 cycles, as shown in Fig. 7.8e.

On the other hand, the electron transfer over the insulating Li$_2$O$_2$ products become complicated, triggering a large ohmic loss and a subsequent rise in the overpotential during charging. Consequently, to conserve catalysts and avoid deactivation after discharging tailored design and successful positioning of catalyst materials on the oxygen electrode are essential for the extra upgrade of Li–O$_2$ battery systems. Ryu et al. designed catalytic membrane via immobilization of Pd nanoparticles on a polyacrylonitrile (PAN) nanofiber and the schematic diagram of a Li-O$_2$ Cell retaining a mesoporous catalytic polymer membrane shown in Fig. 7.9a. SEM images of the Pd/PAN membrane in which Pd nanoparticles agglomerates with a size between 30 and 100 nm on Pan nanofibers represented in Fig. 7.9b cycle performance of as-prepared sample with and without catalytic membrane was tested between 4.35 and 2.35 at 500 mA/g$_{carbon}$ with a limited capacity of 1000 mA/g$_{carbon}$. As a result, a

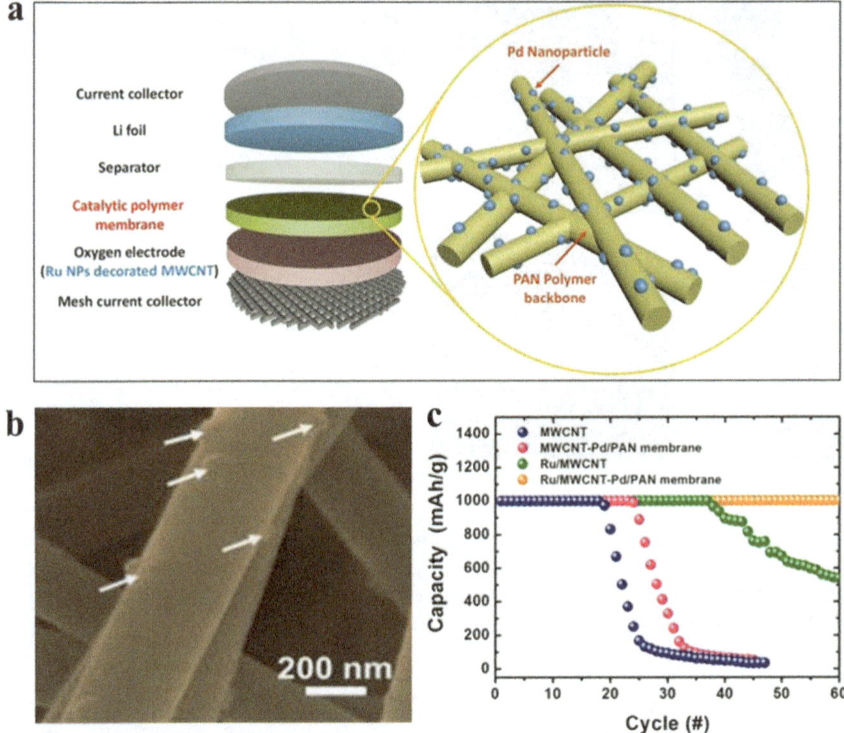

Fig. 7.9 **a** Schematic diagram of a Li–O_2 Cell using a mesoporous catalytic polymer membrane. **b** SEM images of the Pd/PAN membrane. **c** Cycling performance of as-prepared sample with and without catalytic membrane. Reproduced from Ref. [37]. Copyright 2014, American Chemical Society

Ru/MWCNT electrode with a polymer catalytic membrane holds a capacity value of 1000 mAh/g_{carbon} for 60 cycles without any capacity loss which indicates that the kinetics has been improved by Pd NPs [37] demonstrated in Fig. 7.9c.

By the way, Kang et al. reported the polymorphic difference of TiO_2 nanofiber catalysts from anatase to rutile enabled Li–O_2 cells to have elevated round-trip productivity and lesser overpotential resulted by a better cyclic retention as well decrease over the potential that credited to the smaller Li^+ chemisorption energy and bandgap of the rutile phase compared with the anatase phase [38]. We also summarized the in Table 7.1 with the recently reported electrospun nanofibers-based materials, process techniques and significant electrochemical properties.

Table 7.1 Illustrates the recently reported electrospun nanofibers-based materials for LABs

Materials	precursor	Heat treatment	LAB performance	References
Fe/Fe$_3$C CNFs	FeC$_2$O$_4$ · 2H$_2$O/PAN/DMF	1st: 80 °C for 24 h 2nd: 260 °C for 3 h in air	0.6 A h g^{-1} for 41 cycles at 300 mA g^{-1}	[36]
FeNO-CNT-CNFFs	(Fe(Acac)$_3$)/Co(Acac)$_2$/Ni(Acac)$_2$/PAN/DMF	700, 800, and 900 °C for 2 h in N$_2$	30mAh/(g of active catalyst) for 21 cycles	[35]
TiO$_2$ NFs (anatase vs. rutile)	Ti[OCH(CH$_3$)$_2$]$_4$/PVP/ AcOH, EtOH	1st step: 450 °C in air 2nd step: 750 °C in air	1 A h $^{-1}$ for 20 cycles at 200 mA g^{-1}	[38]
CoMn2O4 tube-in-tube NFs	Co(ac)2, Mn(ac)$_2$/PAN/DMF	600 °C for 2 h in air	1 A h g^{-1} for 115 cycles at 200 mA g^{-1}	[39]
δ-MnO2 tubular NFs	KMnO$_4$/CNFs (PAN, DMF)/ water	CNFs: 250 °C for 0.5 h in air & 900 °C for 2 h in Ar	0.5 A h g^{-1} for 21 cycles at 100 mA g^{-1}	[40]
MnO$_2$@NiCo$_2$O$_4$	Co(NO$_3$)$_2$ · 6H$_2$O,/ Ni(NO$_3$)$_2$·6H$_2$O/CO(NH$_2$)$_2$	300 °C for 2 h in air	500 mAh g^{-1} for 168 cycles at 400 mA g^{-1}	[15]
Co$_3$O$_4$ NFs/GNF	Co(ac)$_2$ · 4H$_2$O/PVP/DMF	600 °C for 2 h in air	1 A h g^{-1} for 80 cycles at 200 mA g^{-1}	[28]
Mn$_3$O$_4$/carbon composite	Mn(OAc)$_2$/PAN/DMF	1st: 280 °C for in Ar 2nd 500 °C for in Ar	-	[29]
La0.75Sr0.25MnO3 NTs	La(NO$_3$)$_3$ · 6H$_2$O, Mn(ac)$_2$· 4H$_2$O, Sr(NO$_3$)$_2$/PVP/DMF	600 °C for 1 h in air	1 A h g^{-1} for 124 cycles at 0.15 mA cm^{-1}	[41]
Pd/CNFs	Pd(CH$_3$COO)$_2$/PAN/DMF	1st: 280 °C for 1 h in Ar 2nd 1000 °C for 1 h in Ar	200 mAh g^{-1} at 20 mA g^{-1}	[33]
Ru/MWCNT-Pd/PAN NF membrane	RuCl$_3$/PAN/DMF	–	1 A h g^{-1} for 60 cycles at 500 mA g^{-1}	[37]
Co4N/CNF	Co(NO$_3$)$_2$ · 6H$_2$O/PAN/DMF	1st: 250 °C for 2 h in Ar 2nd: 900 °C for 2 h in Ar	Limited capacity of 1000 mAh g^{-1} at 200 mA g^{-1}	[31]

References

1. K.-N. Jung, J. Kim, Y. Yamauchi, M.-S. Park, J.-W. Lee, J.H. Kim, Rechargeable lithium–air batteries: a perspective on the development of oxygen electrodes. J. Mater. Chem. A **4**(37), 14050–14068 (2016)
2. J.S. Lee, S. Tai Kim, R. Cao, N.S. Choi, M. Liu, K.T. Lee, J. Cho, Metal–air batteries with high energy density: Li–air versus Zn–air. Adv. Energy Mater. **1**(1), 34–50 (2011)
3. T. Ogasawara, A. Débart, M. Holzapfel, P. Novák, P.G. Bruce, Rechargeable Li2O2 electrode for lithium batteries. J. Am. Chem. Soc. **128**(4), 1390–1393 (2006)
4. C.O. Laoire, S. Mukerjee, K. Abraham, E.J. Plichta, M.A. Hendrickson, Elucidating the mechanism of oxygen reduction for lithium-air battery applications. J. Phys. Chem. C **113**(46), 20127–20134 (2009)
5. C.O. Laoire, S. Mukerjee, K. Abraham, E.J. Plichta, M.A. Hendrickson, Influence of non-aqueous solvents on the electrochemistry of oxygen in the rechargeable lithium– air battery. J. Phys. Chem. C **114**(19), 9178–9186 (2010)
6. M. Winter, J.O. Besenhard, M.E. Spahr, P. Novak, Insertion electrode materials for rechargeable lithium batteries. Adv. Mater. **10**(10), 725–763 (1998)
7. E. Peled, The electrochemical behavior of alkali and alkaline earth metals in non-aqueous battery systems—the solid electrolyte interphase model. J. Electrochem. Soc. **126**(12), 2047 (1979)
8. M. Eswaran, N. Munichandraiah, L. Scanlon, High capacity Li–O2 cell and electrochemical impedance spectroscopy study. Electrochem. Solid-State Lett. **13**(9), A121–A124 (2010)
9. D. Zhang, Z. Fu, Z. Wei, T. Huang, A. Yu, Polarization of oxygen electrode in rechargeable lithium oxygen batteries. J. Electrochem. Soc. **157**(3), A362 (2010)
10. G. Zhang, J. Zheng, R. Liang, C. Zhang, B. Wang, M.a. Hendrickson, E. Plichta, Lithium–air batteries using SWNT/CNF buckypapers as air electrodes. J. Electrochem. Soc. **157**(8), A953–A956 (2010)
11. X.-H. Yang, P. He, Y.-Y. Xia, Preparation of mesocellular carbon foam and its application for lithium/oxygen battery. Electrochem. Commun. **11**(6), 1127–1130 (2009)
12. A. Débart, J. Bao, G. Armstrong, P.G. Bruce, An O2 cathode for rechargeable lithium batteries: the effect of a catalyst. J. Power Sources **174**(2), 1177–1182 (2007)
13. L. Huang, Y. Mao, G. Wang, X. Xia, J. Xie, S. Zhang, G. Du, G. Cao, X. Zhao, Ru-decorated knitted Co 3 O 4 nanowires as a robust carbon/binder-free catalytic cathode for lithium–oxygen batteries. New J. Chem. **40**(8), 6812–6818 (2016)
14. W. Wan, X. Zhu, X. He, Y. Wang, Y. Yan, Y. Wu, Z. Lü, Nanoarchitectured CNTs-grafted graphene foam with hierarchical pores as a binder-free cathode for lithium-oxygen batteries. J. Electrochem. Soc. **165**(9), A1741–A1745 (2018)
15. C. Tang, Y. Mao, J. Xie, Z. Chen, J. Tu, G. Cao, X. Zhao, NiCo 2 O 4/MnO 2 core/shell arrays as a binder-free catalytic cathode for high-performance lithium–oxygen cells. Inorg. Chem. Front. **5**(7), 1707–1713 (2018)
16. H. Xue, X. Mu, J. Tang, X. Fan, H. Gong, T. Wang, J. He, Y. Yamauchi, A nickel cobaltate nanoparticle-decorated hierarchical porous N-doped carbon nanofiber film as a binder-free self-supported cathode for non-aqueous Li–O 2 batteries. J. Mater. Chem. A **4**(23), 9106–9112 (2016)
17. R. Singhal, V. Kalra, Binder-free hierarchically-porous carbon nanofibers decorated with cobalt nanoparticles as efficient cathodes for lithium–oxygen batteries. RSC Adv. **6**(105), 103072–103080 (2016)
18. M. He, P. Zhang, S. Xu, X. Yan, Morphology engineering of Co3O4 nanoarrays as free-standing catalysts for lithium–oxygen batteries. ACS Appl. Mater. Interfaces **8**(36), 23713–23720 (2016)
19. Y. Cui, Z. Wen, Y. Liu, A free-standing-type design for cathodes of rechargeable Li–O 2 batteries. Energy Environ. Sci. **4**(11), 4727–4734 (2011)
20. M.J. Song, M.W. Shin, Fabrication and characterization of carbon nanofiber@ mesoporous carbon core-shell composite for the Li-air battery. Appl. Surf. Sci. **320**, 435–440 (2014)

21. Y. Cao, H. Lu, Q. Hong, J. Bai, J. Wang, X. Li, Co decorated N-doped porous carbon nanofibers as a free-standing cathode for Li-O2 battery: Emphasis on seamlessly continuously hierarchical 3D nano-architecture networks. J. Power Sources **368**, 78–87 (2017)

22. S.M. Crespiera, D. Amantia, E. Knipping, C. Aucher, L. Aubouy, J. Amici, J. Zeng, U. Zubair, C. Francia, S. Bodoardo, Cobalt-doped mesoporous carbon nanofibres as free-standing cathodes for lithium–oxygen batteries. J. Appl. Electrochem. **47**(4), 497–506 (2017)

23. J. Huang, B. Zhang, Y.Y. Xie, W.W.K. Lye, Z.-L. Xu, S. Abouali, M.A. Garakani, J.-Q. Huang, T.-Y. Zhang, B. Huang, Electrospun graphitic carbon nanofibers with in-situ encapsulated Co–Ni nanoparticles as free-standing electrodes for Li–O2 batteries. Carbon **100**, 329–336 (2016)

24. H. Nie, C. Xu, W. Zhou, B. Wu, X. Li, T. Liu, H. Zhang, Free-standing thin webs of activated carbon nanofibers by electrospinning for rechargeable Li–O2 batteries. ACS Appl. Mater. Interfaces **8**(3), 1937–1942 (2016)

25. L. Wang, X. Zhao, Y. Lu, M. Xu, D. Zhang, R.S. Ruoff, K.J. Stevenson, J.B. Goodenough, CoMn2O4 spinel nanoparticles grown on graphene as bifunctional catalyst for lithium-air batteries. J. Electrochem. Soc. **158**(12), A1379–A1382 (2011)

26. J. Li, N. Wang, Y. Zhao, Y. Ding, L. Guan, MnO2 nanoflakes coated on multi-walled carbon nanotubes for rechargeable lithium-air batteries. Electrochem. Commun. **13**(7), 698–700 (2011)

27. Y. Liang, Y. Li, H. Wang, J. Zhou, J. Wang, T. Regier, H. Dai, Co 3 O 4 nanocrystals on graphene as a synergistic catalyst for oxygen reduction reaction. Nat. Mater. **10**(10), 780–786 (2011)

28. W.-H. Ryu, T.-H. Yoon, S.H. Song, S. Jeon, Y.-J. Park, I.-D. Kim, Bifunctional composite catalysts using Co3O4 nanofibers immobilized on nonoxidized graphene nanoflakes for high-capacity and long-cycle Li–O2 batteries. Nano Lett. **13**(9), 4190–4197 (2013)

29. K.-N. Jung, J.-I. Lee, S. Yoon, S.-H. Yeon, W. Chang, K.-H. Shin, J.-W. Lee, Manganese oxide/carbon composite nanofibers: electrospinning preparation and application as a bifunctional cathode for rechargeable lithium–oxygen batteries. J. Mater. Chem. **22**(41), 21845–21848 (2012)

30. K.T. Alali, J. Liu, Q. Liu, R. Li, K. Aljebawi, J. Wang, Grown carbon nanotubes on electrospun carbon nanofibers as a 3D carbon nanomaterial for high energy storage performance. Chem. Select **4**(19), 5437–5458 (2019)

31. K.R. Yoon, K. Shin, J. Park, S.-H. Cho, C. Kim, J.-W. Jung, J.Y. Cheong, H.R. Byon, H.M. Lee, I.-D. Kim, Brush-like cobalt nitride anchored carbon nanofiber membrane: current collector-catalyst integrated cathode for long cycle Li–O2 batteries. ACS Nano **12**(1), 128–139 (2018)

32. Y.-C. Lu, Z. Xu, H.A. Gasteiger, S. Chen, K. Hamad-Schifferli, Y. Shao-Horn, Platinum—gold nanoparticles: a highly active bifunctional electrocatalyst for rechargeable lithium— air batteries. J. Am. Chem. Soc. **132**(35), 12170–12171 (2010)

33. S.M. Crespiera, D. Amantia, E. Knipping, C. Aucher, L. Aubouy, J. Amici, J. Zeng, C. Francia, S. Bodoardo, Electrospun Pd-doped mesoporous carbon nano fibres as catalysts for rechargeable Li–O 2 batteries. RSC Adv. **6**(62), 57335–57345 (2016)

34. H. Park, Y. Kim, S.-D. Kim, D. Seo, S.E. Shim, S.-H. Baeck, An Investigation of the electrochemical properties and performance of electrospun carbon nanofibers for rechargeable lithium-air batteries. J. Nanosci. Nanotechnol. **17**(11), 8175–8179 (2017)

35. D. Ji, S. Peng, D. Safanama, H. Yu, L. Li, G. Yang, X. Qin, M. Srinivasan, S. Adams, S. Ramakrishna, Design of 3-dimensional hierarchical architectures of carbon and highly active transition metals (Fe Co, Ni) as bifunctional oxygen catalysts for hybrid lithium–air batteries. Chem. Mater. **29**(4), 1665–1675 (2017)

36. J. Li, M. Zou, L. Chen, Z. Huang, L. Guan, An efficient bifunctional catalyst of Fe/Fe 3 C carbon nanofibers for rechargeable Li–O 2 batteries. J. Mater. Chem. A **2**(27), 10634–10638 (2014)

37. W.-H. Ryu, F.S. Gittleson, M. Schwab, T. Goh, A.D. Taylor, A mesoporous catalytic membrane architecture for lithium–oxygen battery systems. Nano Lett. **15**(1), 434–441 (2015)

38. S.H. Kang, K. Song, J. Jung, M.R. Jo, Y.-M. Kang, Polymorphism-induced catalysis difference of TiO 2 nanofibers for rechargeable Li–O 2 batteries. J. Mater. Chem. A **2**(46), 19660–19664 (2014)
39. S. Peng, L. Li, Y. Hu, M. Srinivasan, F. Cheng, J. Chen, S. Ramakrishna, Fabrication of spinel one-dimensional architectures by single-spinneret electrospinning for energy storage applications. ACS Nano **9**(2), 1945–1954 (2015)
40. P. Zhang, D. Sun, M. He, J. Lang, S. Xu, X. Yan, Synthesis of porous δ-MnO2 submicron tubes as highly efficient electrocatalyst for rechargeable Li–O2 batteries. Chemsuschem **8**(11), 1972–1979 (2015)
41. J.J. Xu, D. Xu, Z.L. Wang, H.G. Wang, L.L. Zhang, X.B. Zhang, Synthesis of perovskite-based porous La0. 75Sr0. 25MnO3 nanotubes as a highly efficient electrocatalyst for rechargeable lithium–oxygen batteries. Angewandte Chemie Int. Ed. **52**(14), 3887–3890 (2013)

Chapter 8
Summary and Outlook

Batteries dignified to show a mounting character as the primary power supplies for the electric vehicles, portable electronic devices and power tools. So far, numerous materials being investigated as advanced electrode materials, including carbon, metal oxides, nitrides, carbides, and their many derivatives. Meanwhile, the materials used may be distinct; there is an indisputable agreement that novel, low-cost, active materials with engineered nanostructures are vital in solving significant improvements in implementation. Electrospinning process is efficient and versatile of fabricating materials with advantages of very high ratio of surface area to volume, definable porosity and direct scalability for mass production for many applications, particularly in the fabrication energy storage and conversion.

Besides with many other paths of research understudy, 1D nanostructured materials are expected to meaningfully impact to meeting some of these challenges as the structural control of nanomaterials can be as valuable as the composition of the materials themselves. Due to the unique characteristics of electrospun nanofibers such as superior electrical conductivity, persistent mechanical properties, high surface area, versatile, cost-effective and straightforward assembly process, it can extend the huge potential for application in electrochemical energy storage devices, like metal-ion batteries (LIBs, NIBs, KIBs), metal-sulfur battery (Li–S) and metal-air batteries (LABs and ZABs). Even though many distinct methods are proposed to manufacture carbon fibers, electrospinning is the most inexpensive to generate CNFs with diameters down to tens of nanometers. Moreover, the electrospinning method tolerates a simple combination of metal (oxides) devoid of complex chemical treatments. In contrast, other carbon derivatives like CNTs and graphene, electrospun nanofibers similarly appreciate benefits in both low fabrication price and a huge accomplishment. Many investigations conducted over the past years and proven a decent insight and a scientific basis required to regulate the surface morphologies and structures of electrospun nanofibers-based composites. In fact, a huge number of micro and nanostructures have been supplied, and several of them are demonstrated to be successful in upgrading the electrochemical routine. Regard to metal-ion batteries (Li, Na and

© The Editor(s) (if applicable) and The Author(s), under exclusive license   157
to Springer Nature Singapore Pte Ltd. 2020
S. Peng and P. R. Ilango, *Electrospinning of Nanofibers for Battery Applications*,
https://doi.org/10.1007/978-981-15-1428-9_8

K), numerous nanofiber-based electrodes such as metals, nitrides, phosphides metal oxides, carbon materials, and their hybrid materials of growing interest from both the fundamental and application points which plays a wonderful role in enhancing the performance because the metal-ion diffusion shortened, tolerate the volume alteration during charge/discharge and offer extra room for the storage of metal-ion in nanoscale electrodes. Furthermore, electrospun hollow nanofibers are most effective active electrode materials in advanced battery devices because of their superior electrical conductivity, higher reactivity, and robust mechanical properties.

Also, we focused on reviewing the recent progress in numerous electrospun nanofiber materials for Li–S cells in the field of the cathode, separator, and interlayer materials. By condensing the recent progress in numerous electrospun nanofiber materials for Li–S cells, advanced designs of the cathode, separator, and interlayer materials are the key factors to construct Li–S cells with outstanding electrochemical performance. In addition to these compensations of electrospun materials, integrating nanoscale attaching material for expanding the contact between Li_2S_n species can be an evocative method to encourage the performance in Li–S cells critically. The usage of electrospun nanofiber materials in the cathodes, separators, and interlayers of Li–S cells can be summarized like an electrospun S host material is necessary to present suitable porosity for large S loading, superior conductive framework, and valid utilization without binder and current collector. Recently, future post metal-ion technologies like metal-air, including Zn-air and Li-air batteries, have received considerable attention. These systems feature the electrochemical coupling of a metal negative electrode to an air-breathing positive electrode through a suitable electrolyte. Metal-air batteries are between traditional batteries and fuel cells which have the design features of conventional batteries in which a metal is used as the negative electrode. They also have resemblances to conventional fuel cells in that their porous positive electrode structure involves a nonstop and inexhaustible oxygen supply from the adjacent air as the reactant, making possible very high theoretical energy densities about 2–10 folds higher than those of LIBs. Developing bifunctional electrocatalysts for metal-air to overcome high activation overpotential and sluggish kinetics for both ORR and OER is of great importance.

Therefore, many investigations combine with MOF, metal nanoparticles, metal oxides, 2D materials heteroatom-doped carbon composite nanofibers used as nonprecious and high-performance oxygen electrocatalysts towards ORR and OER, even rechargeable ZABs and LABs. In this book, we reviewed, regarding the fabrication and characteristics of electrospun nanofibers and their hybrid composites with metal, metal oxides, transition metal chalcogenides, metal nitrates metal carbides and other conductive spices. Remarkably, we discussed the effects of various binder-free and freestanding electrospun nanofibers preparation set-ups and unique polymer precursors, as well as handling parameters. Electrospun nanofibers perform many crucial roles, including the formation of conductive networks, electrode materials for Li and Na ion, K-ion storage, scaffold to support various materials as electrocatalysts (OER and ORR catalyst in LABs and ZABs). Along with the constantly heightened production rate and diminished cost, the development of CNF-based nanostructures would hold enormous pledge to recognize the next-generation batteries.

The manufacturer's authorised representative in the EU is Springer
Nature Customer Service Centre GmbH, Europaplatz 3, 69115 Heidelberg,
Germany. If you have any concerns regarding our products, please
contact ProductSafety@springernature.com

Printed and bound by CPI Group (UK) Ltd, Croydon, CR0 4YY

28/11/2025

02007724-0014